Generis

PUBLISHING

AF682508

Genetic diversity of some Sudanese date palm cultivars using ISSR Markers and SDS-PAGE

Manal A. Ibrahim
Gamal E. Osman

Title: Genetic diversity of some Sudanese date palm cultivars using ISSR Markers and SDS-PAGE

ISBN: 979-8-88676-116-0

Author: Manal A. Ibrahim, Gamal E. Osman

Cover image: www.pixabay.com

Publisher: Generis Publishing
Online orders: www.generis-publishing.com
Contact email: info@generis-publishing.com

ACKNOWLEDGEMENT

I am very much grateful to God, who has given me ability to complete my study. I am deeply indebted and wish to express my sincere gratitude to Dr. Manal Abdulla Ibrahim, Botany Department, Faculty of Science and Technology, Omdurman Islamic University, for her invaluable guidance, encouragement, critical comments and constant inspiration throughout the course of this study.

I would like to express my deep gratitude to Biology Department at Imam Abdalruhman bin Faisal University for their help in offering me the resources in running the study. My grateful thanks are also extended to Dr. Adil Al-Issa , Dr. Mahmad Berekaa and Mr. Al-khalifa Ali for their helps in during the practical and data analysis.

I would not forget to remember Dr. Ishaq Fathulrahman, Chemistry Department for his encouragement.

Table of Contents

List of Tables

List of figures

List of abbreviations

abbreviation	Full Name
FAO	Food and Agriculture Organization
µg	microgram
AFLP	Amplified fragment length polymorphism
APS	Ammonium per sulphate
Bp	Base pairs
BPB	Bromo phenol blue
CBB	Coomassie brilliant blue
CTAB	Cetyl trimethylethyl ammonium bromide
DMSO	Dimethyl sulfoxide
DNA	Deoxyribonucleic acid
dNTPs	Deoxyribose nucleotide triphosphate
DW	Distilled water
EDTA	Ethylene diamine tetra acetic acid
EtBr	Ethidium bromide
Etc	Etcetera
Fig.	Figure
gm	Gram
GS	Genetic similarity
HCL	Hydrogen chloride
KSA	Kingdom of Saudi Arabia
ISSR	Inter simple sequence repeats
Kb	Kilo base
kD	Kilo Dalton
M	Molar
ml	Milliliter
min	Minutes
ng	Nanogram
nm	Nanometer
No.	Number
O.D	Optical density
PCR	Polymerase chain reaction
PGRs	Plant growth regulators
pH	Negative log of hydrogen ion concentration
PVP	Poly vinyl pyrrolidone
RAMPO	Random amplified microsatellite polymorphism
RAPD	Random amplification polymorphic DNA
RFLP	Restriction fragment length polymorphism

RNA	Ribonucleic acid
SDS-PAGE	Sodium dodecyl sulphate- polyacrylamide gel electrophoresis
Sr.	Serial
SSR	Simple sequence repeat
STRs	Short tandem repeats
TBE	Tris-borate Ethylene diamine tetra acetic acid
TCA	Tri-chloro acetic acid
TE	Tris- EDTA
TC	Tissue culture
UAE	The United Arab Emirates
UPGMA	Unweighted pairwise group method with arithmetic mean
USA	The United States of America
UV	Ultra violet
VNTRs	Variable number of tandem repeats

ABSTRACT

Assessments of genetic diversity in date palm cultivars facilitate effective selection and shorten the breeding time, two techniques (SDS-PAGE) based protein and Inter-simple sequence repeat (ISSR-PCR) were used to evaluate genetic diversity of five Sudanese date palm cultivars. Pioneer results not recorded in any previous work in the available literature in Sudan were obtained. From eight protein bands and eight ISSR primers 53 polymorphic bands produced, the genetic similarity between cultivars were estimated by Jacquard's similarity index and clustered in UPGMA and NJ and clustering were used to analyze the data set.The results revealed that a range of genetic variation within each of the Sudanese cultivars were exists. The foreign date palm cultivars recently introduced to Sudan groves are closely grouped in one cluster this could be explained by the presence of a common genetic origin among them and cultivars Mishriq and Barhi formed a distinct group suggesting their close relatedness. The most remarkable result of the current study is that some genotypes revealed a unique profile with the primers and the results were reproducible and this can be used for DNA fingerprinting and cultivar identification. Cluster analysis on ISSR and SDS-PAGE data clearly discriminated the high degree of independence between the geographical origin and molecular data was indicated. More researches should be done for date palm conservation through biotechnological research to identify and quantify genetic diversity components in the species, identify and clone genes and gene complexes for biotic and abiotic stresses, and utilize the generated information for future research and development.

CHAPTER ONE INTRODUCTION

1.1 Introduction

The main biological features of tree species include the characteristic long life cycles and the lack of mobility which necessitate a need to withstand exposure to large fluctuation in their environmental conditions. Date palm (Phoenix dactyliferaL.) is an angiosperm that belongs to monocots, and is considered the most important and ancient cultivated species in the Arab world (Elshibli, 2009). It was found in Iraq and Egypt before 3000 BC (Al– Ekidy2000), Date Palm has a major socio-economic importance of its high nutritional value, great yields and its long life span.

Date palms in Sudan has been traditionally grown using old, local varieties, mainly of the dry type, for 3000 years. However, semi dry and soft varieties are also grown in limited areas and numbers. The classification of date palm fruits into dry and soft types, mainly depending on the texture of the ripe fruit, is related to some specific content of water and different form of sugarsas well as the sugars acidity (Elshibli and Korpelainen, 2009b).

The most important indigenous dry and soft varieties known include (Barakawi, Gondaila, Tamoda and Abdel Rahim) and (Mishrig Wad Khateeb, and Mishrig Wad Laggai) respectively. Moreover, a large number of farmer" verities resulting from seeds are also locally grown by farmers and they are named collectively as Jaw indicating that they are seedling varieties. This is in addition to a number of males that are used as source of pollen for hand pollination of the female trees. (Osman and Boulos, 1978).

The weather in Sudan makes northern Sudan an excellent location for date production. Several varieties have been recognized and selected by farmers, then cloned and cultivated for commercial production and local use. It has been reported that in Sudan there are about 400 varieties and strains of date palm, including dry, semi-dry and soft dates (Osman 1984).

The characterization of date palm cultivars is of paramount importance toward the conservation of diversity of date palm. There are thousands of varieties existing, but most of them are either synonyms of one variety beingcultivated in inter- or intra-zones or varietal lines. The characterization of these varieties can be done by evaluating the fruit morphological traits, viz. shape, size, weight, color, and texture (El-Houmaizi et al. 2002; Markhandet al. 2010), and by using some biochemical markers such as isozymes and proteins (Baaziz and Saaidi 1988).

The protein electrophoresis is a powerful tool in elucidating the origin and the evolution of cultivated plants (Ladizinsky and Hymowitz,1979). Therefore, protein pattern analysis by gel electrophoresis has been used in higher plants to study various problems in breeding, genetics, taxonomy, physiology, (Leonardo et al., 2010; Chang et al., 2010; Castillejo et al.,2012; Elshibli and Korpelainen, 2008; Davar et al., 2012) .

DNA markers have proved valuable in crop breeding, especially in studies on genetic diversity and gene mapping. The commonly used polymerase chain reaction (PCR)-based DNA marker systems are random amplifiedpolymorphic DNA (RAPD) ,amplified fragment length polymorphism (AFLP) and more recently simple sequence repeats (SSRs) or microsatellites (Staubetal., 1996;Gupta &Varshney, 2000). The major limitations of these methods are low reproducibility of RAPD, high cost of AFLP and the needto know the flanking sequences to develop species specific primers for SSR polymorphism.

Various molecular techniques have been used for date palm cultivars identification from several countries (Bodian et al. 2012a, b) such as random amplified polymorphic DNA (RAPD) markers for Pakistani and Moroccan varieties (Mirbahar et al. 2014; Sedra et al. 1998), amplified fragment length polymorphism (AFLP) for Egyptian varieties (El-Assar et al. 2005), inter- simple sequence repeats (ISSR) for Iraqi varieties (Khierallah et al. 2014), random amplified microsatellite polymorphism (RAMPO) for Tunisian varieties (Sakka et al. 2004), and microsatellites for Moroccan varieties (Bodian et al. 2012a). However, there are huge technical differences in termsof cost, speed, amount of DNA needed, technical labor, degree of polymorphism, accuracy of genetic linkage estimates, and the statisticalpower of tests (Garcia et al. 2004).

ISSR-PCR is a technique that overcomes most of these limitations (Zietkiewicz et al., 1994; Gupta et al., 1994; Wu et al., 1994; Meyer et al., 1993). It is rapidly being used by the research community in various fields of plant improvement (Godwin et al., 1997). There are few reports on molecular markers from genomic sequences in date palm; in Sudan only genomic SSR and ISSR markers have been identified. The ISSR- PCR technique is useful in areas of genetic diversity, phylogenetic studies, gene tagging, genome mapping and evolutionary biology in a wide range of crop species.The demand for different date palm cultivar are important for designing a sustainable genetic resources conservation strategy, but do not often receive much attention in Sudan.

1.2 Statement of the problem

(1) The potential for date palm cultivation is getting more important with the increasing interest to expand its culture throughout the country south to the latitude 13 N.

(2) Evaluation of different date varieties will allow us to estimate the variability of date varieties.

(3) Increased availability of these markers would aid in the genetic and genomic studies in date palm as they are better tools than RAPD markers because of their co-dominant inheritance, multi-allelic nature, and high reproducibility.

(4) The demand for different date palm varieties are important for designing a sustainable genetic resources conservation strategy, but do not often receive much attention in Sudan.

1.3 Objectives of the Study

(1) To determine the inter varietal genetic polymorphism among those cultivars.

(2) Enlarge the number of molecular marker that is suitable to identify the cultivars under study, to enrich the genetic and genomic.

(3) To asses suitable molecular markers for their future selection and potential breeding by using ISSR (inter - simple sequence repeat) and gel electrophoresis.

CHAPTER TWO REVIEW OF LITERATURE

2.1 Botanical description of the date palm

Date palm (*Phoenix dactylifera* L.) is a dioecious, perennial, diploid, and monocotyledonous plant (Zhao et al., 2012). It contains 18 pairs of chromosomes (2n = 36), where the XX/XY system is used to determine its sex. Its genome has recently been sequenced, showing a length of ~ 671 Mb and including ~ 41,600 genes (Al-Mssallem et al., 2013).

Female trees are cultivated mainly for their nutritive fruits. Although the average economic life of a date palm tree is estimated to be up to 50 years, the tree may stay productive up to 150 years (Chao and Krueger, 2007). It is a tall tree which reaches a height of 15-20 m before it is cut down due to declining yield, increasing difficulty and danger in reaching the top during pollination, bunch management and harvesting (Shamsi & Mazloum zadeh, 2009).

The roots originate from a bulb at the trunk base and have a fasciculated fibrous system similar to a maize plant, with their length reaching 4 m on average and up to 10 m in light soil (Manickavasagan et al., 2012). In a deep loamy soil approximately 85% of date palm roots are distributed in a zone to2 m deep and 2 m on both lateral sides. These roots can withstand wet soilfor months, but if such conditions continue for longer periods they may harmthe roots and thus the fruit production (Zaid & de Wet, 2002).

The trunk of date palm, also called a stipe or stem, is brown in colour and isa single vertical cylinder which can extend to 30 m in length. It is un- branched and borne singly or in clumps by production of offshoots with a single terminal meristem protected by developing leaves and there is no cambium. The trunk is formed after the meristem has reached its full diameter after which there is no addition to the tissues of the columnar trunk or in the number of vascular bundles (Zaid, 2002). It is covered by leaf basesthat are enclosed in fibre which protect the trunk from animals and insects, and in addition reduce loss of water (Manickavasagan et al., 2012; Zaid & deWet, 2002).

The leaves of date palms have a normal life span of 3–7 years and can reachup to 6 m in length and half a meter in width. The leaf narrows at the midrib and towards both leaf ends. The mature trees have approximately one hundred to one hundred and twenty-five leaves and produce ten to twenty- six new leaves annually. The leaf can support the production of date clusters from 1–1.5 kg under good conditions (Manickavasagan et al., 2012; Zaid & de Wet, 2002).

2.2 Nutritional and medicinal value of dates

Date palm is considered as an important constituent of farming systems in dry and semi-arid regions, and is suitable for both small and large scale farming (Khushk et al., 2009). It is one of the fruit trees with highestproduction per hectare (Okorley et al., 2005).

The high nutritional composition, profitability as well as environmental advantages makes date palm an excellent choice for the farmers (AlShahib and Marshall, 2013). Its cultivation is a good alternative to improve the food and economic status of people (Hassan et al., 2006). Similarly Chao and Krueger (2007) reported that date palm tree has numerous usages and economic importance in ecological improvement of the deserts.

On the palate, dates have a fleshy mouth feel and a sweet taste. They provide essential nutrients and are claimed to have benefits in human health (Manick avasagan et al., 2012). Date contains many nutrients such as: carbohydrates, proteins, fat, minerals and vitamins (Al-Qarawi et al., 2004; Barghini et al., 2007). It is good source of high nutritional value food. Indeed it is rich in carbohydrates, dietary fibers, proteins, minerals andvitamin B complex, such as thiamine (B1), riboflavin (B2), niacin (B3), pantothenic (B5), pyridoxine (B6), and folate (B9) (Chao and Krueger, 2007; Al-Harrasi et al., 2014; Siddiq et al., 2013; Eoin, 2016). In more details, carbohydrates forms 70% of date fruit and are mostly fructose and glucose in equal ratio while date proteins are rich in aminoacids that contain acidic side chain but poor in methionine and cysteine, which their side chain composed of sulfur. Minerals in date fruits are calcium, iron, magnesium, selenium, copper, phosphorus, potassium, zinc, sulfur, cobalt, fluorine, manganese, and boron (Chao and Krueger, 2007; Al- Harrasi et al., 2014; Siddiq et al., 2013; Eoin, 2016).

Al-Farsi et al, (2007) pointed out that date by-products can be considered a good source of dietary fiber, total phenolic and an inexpensive source of natural antioxidants and could be used as a functional food ingredient.

The edible part of the date palm tree has been recognized to possess many medicinal properties when consumed alone or in mixture with other medicinal herbs (Tiwari et al., 2013). In recent years, a huge interest in the abundant health promoting properties of date fruits had led to many pharmacological studies (in-vitro and in-vivo) as well as the identification and quantification of different classes of phytochemicals (Chao and Krueger,2007) .

The date fruits are highly nourishing and may have numerous potential health benefits. The protective effects of fruits against chronic diseases are ascribed to bioactive non-nutrients called phytochemicals. Phytochemicals have gained increased interest among several investigators , including clinicians due to their antioxidant activity, cholesterol-lowering properties, and other potential health benefits such as chemoprevention of cancer, prevention of diabetes and cardiovascular diseases (Chao and Krueger, 2007;Tiwari et al., 2013; Al-Harrasi et al., 2014).

2.3 Date palm genetic diversity

Jaradat (2015) defined genetic diversity as the genetic variation between species, subspecies, varieties, populations, or individual clones that can be measured morphologically, physiologically, biochemically, and at the molecular level. Date palm varieties have evolved by thousands of years of seedling selection with desired characteristics and wide range of genetic diversity in fruit quality, and shape. Each variety is derived from a unique single seed, cloned and vegetative propagation (Baaziz et al. 2000; Chao andKrueger 2007).

At the global level, over 5000 date palm varieties exit in date palm growing countries, but sometimes might be synonyms of one variety found in different countries under a different name, but about 10% of them of a commercial importance (Baaziz et al. 2000), . However, each country got its own top elite varieties of commercial value. A Sudanese most famous variety 'Barkawee' is widely cultivated in northern of the country.

There are two genetic diversity concerns when rapid multiplication through tissue culture is considered. First, the propagation of date palms through tissue culture has been accompanied by challenges that affect the level of certainty of true-to-typeness of propagules. However, commercial micropropagation and distribution of geno typically stable varieties have been ongoing since early 1990s (Zaid and de Wet 2002b).The modified or mutated in vitro plants can result in considerable pollution for the traditional, slowly developed varieties, if not recognized and carefully considered. The second concern is that the success to clone is found to be variety -specific (Al-Khayri and Al-Bahrany 2004; Pinker et al. 2009). Similar selectivity among genotypes has been also observed within natural cloning systems, i.e.vegetative propagation, of date palm. For example, in Sudan, it is well- known that the dry type cv. Barakawi produces more offshoots per mother tree than the soft type cv. Mishrig Wad Laggai. The issue of farmers" shift to monoculture is again raised when disease-tolerant varieties are considered, as resistance to bayoud disease has been found to be variety-specific (Baazizet al. 2000).

Reported constraints suggest that large numbers of traditional varieties in many countries may be endangered or diminishing, and it has been suggested that urgent conservation efforts are needed worldwide (Ahmedand Al-Qaradawi, 2009; Baaziz et al. 2000; Chao and Krueger 2007).

2.4. Protein identification, analysis and characterization

Understanding function of the biological systems requires knowledge of their chemical composition. The state-of life of a cell at any given time isdefined by its protein composition, i.e. its proteome. That is why it isnecessary to have a rapid and an efficient tool for characterization of theproteome, (Pinker et al. 2009).

There are different amounts of storage proteins in all plant parts. They play two main roles including nitrogen and energy source and defense against insects and pathogens such as bacteria and fungi (Ehsan pour et al. 2010). Among them seed storage protein analyses helps in identification and characterization of diversity in crop varieties, varieties and their wild varieties and also provides information on phylogenetic relationship of the accessions.

There are several ways of protein identification, however, majority of them are very time consuming procedures. A modern and efficient tool is based on comparison of theoretically derived molecular masses of peptides from known DNA sequences and primary protein structures in database withexperimentally obtained molecular masses of enzymes generated peptides ofseparated proteins. The mass spectrometry based strategy designed for protein identification exhibits good sensitivity and this approach allows evendetailed function analyses.

Separation techniques like one dimensional gel electrophoresis, mass spectrometry (i.e. matrix assisted laser desorption/ionization time of flight mass spectrometry) and bioinformatics (i.e. database searching) became important tools for protein and proteome analysis (Chang et al., 2010).These methods have been successfully applied for identification of proteins in numerous biological systems (2001, Mann et al. 2001, and Chemelik et al.2002).

Gel electrophoresis is the study of the movement of charged molecules in an electric field. Generally cellulose or thin gel made up of their polyacrylamide or a garose is used as support medium. Cellulose is used as support medium for low molecular weight biochemical such as amino acid and carbohydrates where as a garose and polyacrylamide gels are widely used for larger molecules like proteins. Protein pattern analysis by gel electrophoresis has been used in higher plantsto study various problems in breeding, genetics, taxonomy, physiology, (Leonardo et al., 2010; Chang et al., 2010; Castillejo et al., 2012; Elshibliand Korpelainen,

2008; Davar et al., 2012) as it has been appreciated as a biochemical tool for studying the phylogentic relationships (Al-Yahyai and Al-Khanjari, 2008). The phylogenetic analysis based on protein patterns were used to study the genetic relatedness between and among varieties (El- Hady et al., 2010; Abdulla and Gamal, 2010; Haider et al., 2012).

SDS-PAGE is the most widely used analytical method to resolve separate components of a protein mixture. It is almost obligatory to assess the purity of a protein through an electrophoretic method. SDS-PAGE is effectively used to evaluate and characterized the protein patterns of seed storage proteins and to study genetic diversity of the plant species.

Ahmad, (2008) concluded that SDS-PAGE provided valuable informationfor the identification of populations and could be utilized for population and varietal discrimination as well as seed quality tests in true to type seed producing plants.

Thanh et, al :(2006). They used electrophoretic markers of the seed storage proteins for identification of varieties, for screening of pure line to propagateinto new variety with high quality, and to check species identification and evaluation of genetic diversity of the species.

Recently, Munshi and Osman (2010) used sodium dodecyl sulfate polyacrylamide gel electrophoresis (SDSPAGE) to analyze total proteins of fresh leaves for determination of molecular phylogeny of some date palm varieties in Saudi Arabia. The authors reported that no major differences in banding patterns of the protein of the studied varieties were recognized.

Isozymes markers are proteins based produced by gene expression. These proteins can be identified and isolated through the electrophoresis and staining technique. Isozyme the different molecular form of same enzyme that catalyses the same reaction, are proteins, that can be identified through the enzymatic associated coloured reaction. This process can be done on electropherogram and these are the product of various alleles of one or several genes. Isozyme (or isoenzyme) was the first biochemical markers extensively used to study of plant, animals and insects etc. (Mojtaba K, Mehdi R.2015).

During this method crude protein is extracted which further separated through electrophoresis on starch or non-denaturing Poly acrylamide gels electrophoresis (PAGE). Appropriate enzymes are added which is mandatory for the specific activity of each enzyme and for the visualization of isozymes zones. And resulting isozymes banding patterns (zymograms) are used to find out genetic relationships. (Mojtaba K, Mehdi R.2015).

2.5. Molecular markers

Molecular markers are genetic loci that can be easily traced and measured in a population and may be associated with a particular gene or trait of interest. Molecular markers or DNA markers have been in use since past three decades. The first reported use of DNA Polymorphism detection was the construction of a genetic map through Restriction Fragment Length Polymorphism (RFLP) (Botstien et al., 1980)

The development of different molecular markers enables scientists to accurately identify dates variety with their genetic diversity and phylogenetic relationships. Genetic conservation is reliant on understanding the extent, and distribution of the genetic diversity exists in the prevailing germplasm (Jubrael et al. 2005). In addition, due to the accessibility of nuclear (Al-Dous et al. 2011) and chloroplast (Yang et al. 2010; Khan et al. 2011. (Soumaya et al. 2014) genome sequences of date palm, a number of molecular marker systems can be applicable for an effective assessment of the genetic diversity for germplasm conservation and exploitation of true-to- type elite material (Khierallah et al. 2011).

Present day different types of molecular markers, and have different principle and methodologies Randomly Amplified Polymorphic DNA (RAPD) are the first generation of molecular markers and have been used alone or in combination with other molecular markers for diversity analysis and variety identification in date palm (Al-Khalifa, 2006; Rani et al., 2007; Marsafari and Mehrab, 2013; Mirbahar et al., 2014; Ahmad and AlHadidi, 2014; Yusuf et al., 2015). RAPD Markers is a PCR-based marker technique, requires only small amount of genomic DNA (usually 10 bp), and is a quite inexpensive way as it does not require blotting and radioactive material like other molecular techniques (Mirbahar et al. 2014; Emoghene et al. 2015). Several studies conducted using RAPD markers to detect genetic diversity and polymorphism in date varieties.

(Al-Khalifah and Askari 2013) Random amplified polymorphic DNA (RAPD) fingerprints have been used to study the genetic diversity andidentify 13 date palm varieties in Saudi Arabia. In this study, 140 RAPD primers were screened, of which 37 primers revealed polymorphism among varieties. The number of polymorphic bands per primer varied between 2 and 17, with a mean of 4 major bands per primer.

The use of RAPD markers is very simple and economical but these markers are dominant in nature and their results are less reproducible therefore these are less reliable (Semagn et al., 2006).

Amplified Fragment Length Polymorphism (AFLP) Markers, the principle behind this technique is the PCR amplification of a selected subset of all fragments produced by restriction digest of genomic DNA. The AFLP technology is sensitive enough to detect low levels of variation, allowing us to discriminate between highly related varieties (Diaz et al., 2003).A large number of studies using the AFLP as molecular markers have been reported in the literature.

Elhoumaizi et al. (2006) confirmed that „Medjool" is a landrace variety in Morocco by AFLP analyses of „Medjool" accessions from Morocco, Egypt, and USA. They found that „Medjool" is not genetically uniform and most probably evolved by natural selection from a mixture of genotypes under the grown environmental condition.

Microsatellites or Simple Sequence Repeats (SSRs) are segments of repeated DNA sequence in higher eukaryotic genome. Usually comprised of 1-6nucleotide repeats. They can detect length variation with the help of Polymerase Chain Reaction (PCR). These markers are popular for the analysis of plant diversity because they are PCR based co-dominant markers which show high allelic diversity at different loci (Elmeer et al., 2011). Simple sequence repeats are abundantly present in the higher eukaryotic genomes and are highly polymorphic than other genetic markers. These are helpful in the identification of varieties, analysis of pedigree, germplasm characterization and genetic mapping studies (Billote et al., 2004).

The first SSR markers published for date palm were constructed by Billotte et al. (2004). This microsatellite library was rich in (GA) n repeats and characterized 16 SSR loci in date palm.

Hamweih et al. (2010) developed 1000 microsatellite markers through the date palm genome. They found dinucleotide repeats to be the most frequent (52442 motifs) type of microsatellite repeats with GA being the most common dinucleotide (48.7%) among them. They suggested their use in genetic study and diversity analysis of date palm.

These markers have been used for the assessment of genetic diversity of datepalm varieties in Qatar (Elmeer et al., 2011), Pakistan (Faqir et al., 2016)and Iran (Khierallah et al., 2011). In all these studies SSR markers groupedcthe studied varieties into different clusters but the identification of individualvarieties was still not possible.

Maryam et al. (2016) used 12 SSR primers to assess 30 date palm samples including 8 males and 6 females. These primers also classified the studied date palm samples into male and female plants. SSR markers used in this study were found to be useful in sex discrimination of the date palm. Also Al-Faifi et al. (2016) developed 93 microsatellite markers and detected high genetic diversity among 32 Saudi date palm varieties by the use of SSR markers.

Al-Najm et al. (2017) used SSR markers developed by Billotte et al. (2004)and Akkak et al. (2009) for analysis of molecular diversity and populationstructure of 60 date palm varieties grown in Australia. They found thatAustralian germplasm has huge diversity providing basis for date palmimprovement, increase in production and conservation of genetic resources. Variation in these SSR markers will help in the development of markersassociated with biochemical or physiological functions. Filho et al (2017)has identified 1555726 microsatellites in date palm and oil palm that bringsopportunity for development of molecular markers to be applied in breedingprograms for selection of superior genotypes.

Inter Simple Sequence Repeat (ISSR) markers; this marker includes DNA segments amplification available in between two identical microsatelliterepeat regions oriented in an opposite direction. This uses SSR as primers in a single primer PCR reaction targeting multiple genomic loci. For ISSRs, these microsatellite repeats used as primers. The used primers can be unanchored (Gorji AM, et al. 2011), (Luo C, et al. 2010), Varshney RK, etal. 2005), or mostly anchored at 3` or 5` end (Welsh J, et al. 1990).

ISSRs use longer primers (15–30 mers), which allow the further use of high Tm resulting to higher stringency. Tm depends on the (GC) content of the used primer and that vary from 45 to 65°C. These amplicons are 200–2000 bp long and can be detected by both polyacrylamide gel electrophoresis and agarose gel electrophoresis. ISSRs need no prior primer sequence information (Harushima Y, et al. 1998) this technique is quick, simple, and radioactivity use is not mandatory. Usually, ISSR markers show high polymorphism (Joshi SP et al. 2000).

ISSRs markers are powerful tool to study the inter- and intra-specific genetic variations in date palm and an easy approach with highly reproducible and multiple genomic loci target ability Several works have been done so far on assessing genetic diversity using ISSR markers in date palm varieties.

Ahmad and Al-Hadidi (2014) used eighteen ISSR markers for the evaluationof genetic diversity and relationship among fifteen varieties from Qatar; theyfound that although the varieties were different in their agronomic traits genetically they were interrelated. Hammadi et al. (2012) found that fruit consistency which is an important characteristic of date fruit has association with genetic markers because clustering based on fruit consistency is in accordance with clustering by microsatellite markers.

Karim et al. (2010) used seven ISSR primers to estimate genetic diversity among 10 accessions of date palm varieties in Tunisia and suggested use of more primers to clear-cut among date palm varieties. Also Zehdi et al. (2004b) and Karim et al. (2010) obtained useful information regarding phylogenetic relations among Tunisian date varieties using ISSR markers.

In addition , Moghaieb et al. (2010) in an assessment using RAPD and ISSR markers for identification of genetic diversity and sexuality determination of six male and female date genotypes in Qatar found high polymorphism by using ISSR markers (73%) compared with RAPD (60.2%).

Abdullah and Gamal (2010) applied three types of markers such as protein RAPD PCR and ISSR on four important varieties in Saudi Arabia i.e Med 3002b1, Sugaylb1, Halasb1 and Sukkari b. Inter varietal variation was investigated using five RAPD and five ISSR markers were also applied to assess the genetic polymorphism. Most of the varieties had the narrowgenetic diversity as already expected. The result of the analysis can be used for the selection of the possible parents to generate mapping population.

Single nucleotide polymorphism (SNPs), this marker is generally based on single-base pair position differences in genome sequence. Single nucleotide exchanges in genomes are in large numbers; hence these SNPs markerproduces a great marker density. Single nucleotide polymorphisms (SNPs) are the most potential markers to create a very high-density genetic map of plants. A group of date palm researchers has been employed the application of SNRs in studying the date palm genome sequences.

Mathew et al. (2015) conducted genotyping-by-sequencing from fruit and leaf samples of 70 female varieties of date palm from different countries andgenerated genome-wide genotypic data for 13,000–65,000 SNPs. They found that date palm varieties segregate into two main regions of shared genetic background from North Africa and Arabian Gulf.

Recently, Hazzouri et al. (2015) achieved a complete catalog of about seven million SNPs from 62 varieties of date palm based on whole genome resequencing. The findings indicated the major genetic division between North Africa and the Middle East/South Asian date palms, with evidence of admixture in varieties from Egypt and Sudan.

CHAPTER THREE MATERIALS AND METHODS

3.1. Materials

3.1.1 Plant materials. See appendix 3.1.1

For protein and deoxyribonucleic acid (DNA) analysis, green and yellow leaflets samples from 5 female date palms, comprising 5 varieties, were collected from Selate farm in Khartoum. The (Barhi, Khenaizi, Khadhrawi, and Sugay), have been recently introduced from Kingdom of Saudi Arabia (Sugay), United Arab Emirates (Khonaizi), Iraq (Khadhrawi and Barhi) and Mishriq Wad Laggai which is domesticated in Sudan, were collected and transferred in Liquid nitrogen (-190°C).

3.1.2 Reagents and chemicals for protein analysis. See appendix 3.1.2

3.1.3 Reagents and chemicals for DNA analysis. See appendix 3.1.3

3.1.4 DNA extraction Buffer. See appendix 3.1.4

3.1.5 DNA extraction instruments and glassware. See appendix 3.1.5.

3.2.1 Method for protein analysis

3.2.1.1 Protein extraction

Protein extracted method according to Al-Helal (1994) with some modifications The samples were ground in a mortar by liquid nitrogen, then the proteins extracted by homogenizing 50 mg of ground leaflet samples by half volume stainless steel beads with 0.7 mL extracting buffer at 9/4 min., followed by 10/2 min. by using of (Bullet Blender Homogenizer), then incubated overnight at 4°C. The crude extract was vortexed by using of (VELP vortex mixer), then centrifuged at 12500 rpm for 10 min by using of (Eppendorf centrifuge 5424), the supernatant was moved to new tubes, whilediscarded the debris.

3.2.1.2 Protein precipitation and re-suspending

I used Methanol and Chloroform precipitation methanol, from the method described by Wessel, and Fluegge, (1984) with some modifications , To 150μL of the supernatant 600 μL methanol was added, vortexes, next 150 μL chloroform was added vortexes , centrifuged at 10000 rpm for 1 min., then top aqueous layer was removed , whole proteins was between layers ,600 μLmethanol was added , vortexes , centrifuged at 10000 rpm for 1 min , then supernatant was removed without disturbing pellet , precipitated protein was dissolved in 150 μL of extracting buffer.

3.3.1. Electrophoresis process

3.3.2. Preparation of polyacrylamide gel. See appendix 3.3.2.

According to Laemmli (1970) with some modifications; I used 12.5% resolving gel , The Separating gel mixture was degassed under vacuum for about 30 seconds and then Tetramethyl ethylenediamine (TEMED) wasadded to the solution with gentle swirling to ensure even mixing, thentransferred to the gel cassette by using micropipette. To prevent the upper surface of the separating gel from the air, saturated butanol solution was overlaid. The gel was then left to polymerize for 20-25 minutes.

After that, stacking gel solution was prepared and degassed before use. Overlaid butanol solution was poured off before pouring of stacking gel. Opened surface of polymerized separating gel was washed by using stackinggel solution. Then stacking gel solution was added to the gel cassette untilthe solution reaches the cut away (notched) edge of the gel plate.

Immediately comb (2 cm depth and 14 cm length) was placed to form wells in stacking gel, and left to polymerize. This will take about 20 min. To facilitate loading of samples, well positions were marked by marker pen. Comb was removed carefully from the stacking gel and the cassette was assembled in the electrophoresis tank. Electrophoresis tank was filled to marked level with electrophoresis buffer.

The sample (10 μL) was loaded, in each well. And 10 μL of protein molecular weight marker (6.5 kD-205 kd) using loading guide and Hamilton syringe. The electrophoresis was carried out by using of (Bio-Rad, Broka
0.75 mm) mini electrophoresis system with. (Bio-Rad Power Pac Basic) at 100-150 Volts. The current was put off and the power supply was disconnected when the PBS line reached the lower part of the gel.

3.3.3. Staining and distaining. See appendix 3.3.3.

The gel was washed with 100 mL of (GFS) for 1 hour, at room temperature, and then covered with 100 mL of (GWS) for overnight with agitation at room temperature, and then aspire the (GWS).

The gel covered with 100 mL of Coomassie brilliant blue R-250, Staining was carried out with shaking for a minimum of 2 h to overnight. The stained gel was then placed in the destining solution until the background become transparent.

3.3.4. Bands scoring and Data analysis

After separating protein products by Sodium Dodecyl Sulphate– Polyacrylamide Gel Electrophoresis agarose (SDS-PAGE), the gels were visualized using a Photo Print (Vilber Lourmat, France) imaging system.The protein bands were scored as the presence or absence of protein bands. Then computed with SPSS-10 program to produce a genetic distance matrix using Dice similarity coefficients19. A dendrogram was generated by clusteranalysis using unweight pair group method of the arithmetic averages (UPGMA).

3.4.1 DNA extraction method. See appendix 3.4.1.

3.4.2 Qualitative and Quantitative Analyses of Extracted DNA

DNA yield was measured using a UV-visible spectrophotometer (PerkinElmer, Waltham, MA, USA) at 260nm. DNA purity was determined by calculating the absorbance ratio at A260/280nm. (Wilson K. and Walker
J. 2005). For quality and yield assessments, electrophoresis was performed for all DNA samples on 0.8% a garose gels that were stained with ethidium bromide; the bands were observed and compared with a known standard lambda DNA marker sample.

3.5.1. ISSR-PCR

3.5.1.Primers Selection: three types of primers, produced by (Sigma Aldrich, Banglore) , six 3 anchored dinucleotides repeat primer (AG)10C, (AG)10T, (CT)10A, (CT)10T, (CT)10G, and (CA)8GT, one non anchored
tetra-nucleotides repeat primer (GACA)4, and one non anchored tri- nucleotides (CAG)5. Which were preselected for their performance with date palm DNA were tested to arrive at the primitive primer which gives descriptive segments (polymorphism).Table: 3.5.1.

3.5.2 The polymerase chain reaction (PCR) Optimization

According to Williams et al 1990, the polymerase chain reaction (PCR) mixture 25 µl consisted of 2 µl of total genomic DNA, 12.5 µl of Ampli TaqGold 360 Mastermix (Applied Biosystems), 2 µl (5 pmol/ µl) of each primers and 8.5 µl of nuclease free water.

Amplification took place in DNA amplification Thermocycler (Biorad, icycler) amplification thermocycler, is programmed as adenaturation step of 4 min at 94

°C. Following by 35 cycles compose of 30 seconds at 94 °C, for 30 seconds at annealing temperature and 3 minutes at 72 °.a final extensionof 72 °C for 5 minutes, and hold at 4°C .

No	Primer	Sequence 5′--- 3′	Annealing Temp. 0C/Sec
1	DPISR1	CAGCAGCAGCAGCAG	50.9 / 57.8
2	DPISR2	CACACACACACACACAGT	50.9 / 57.8
3	DPISR3	GACAGACAGACAGACA	48 / 47.5
4	DPISR4	AGAGAGAGAGAGAGAGAGC	50.9 / 57.8
5	DPISR5	AGAGAGAGAGAGAGAGAGT	50.9 / 57.8
6	DPISR6	CTCTCTCTCTCTCTCTCTA	50.9 / 57.8
7	DPISR7	CTCTCTCTCTCTCTCTCTT	50.9 / 57.8
8	DPISR8	CTCTCTCTCTCTCTCTCTG	50.9 / 57.8

Table 3.5.1: the primers used in ISSR analysis

3.5.3. A garose gel Electrophoresis

Amplification products were electrophoresed on 1.8% agarose gel (Sigma)in 100 mil 0.5XTBE buffer see appendix 3.1.3. The gel was run at 120V constant voltages for 45 minutes. The 100 bp standard DNA molecular weight marker (ladder) (Sigma Aldrich, Banglore) was run along with the samples to compare the molecular weight of amplified products. Gels were stained with 0.5 μg/mL ethidium bromide for 15 min.

3.5.4 Visualization and analysis of PCR products

Visualization of amplification products and data analysis reproducible bands visualized on a garose gels 1.8% were scored using a binary code in a data matrix (1 and 0) for their present and absent respectively for the eight primers. Fragments with same mobility were considered as identical, irrespective of the intensity of the fragment. Data were then computed with SPSS-10 program to produce a genetic distance matrix using Dice similarity coefficients19. A dendrogram was generated by cluster analysis usingunweight pair group method of the arithmetic averages (UPGMA).

CHAPTER FOUR RESULTS AND DISCUSSION

In the present study the level of genetic diversity among five date palm varieties Khenaizi, Sugay, Mishriq, Khadhrawi, and Barhi. Were estimated using two techniques, electrophoretic protein patterns technique and inter simple sequence repeat ISSR- PCR-based.

4.1 Protein pattern analysis

The total soluble proteins of the studied varieties were found to be composed of a total of 8 bands, two bands were common between allvarieties at 15, 25 kda, the number of protein bands ranged between 7 bands in Sugay and Khadhrawi, to 3 in Mishriq, which scored the lowest number of bands due to low concentration of protein. (Comas et al. 1979; Sammour and Sharaf eldin-1989; Vas et al. 2004). The value of polymorphic percentage was 75% which considered as higher percentage. (Fig 4.1).

KHZ SUG MSH KHD BAH

Mw.kda 225

150

100

75

50

37

25

15

12

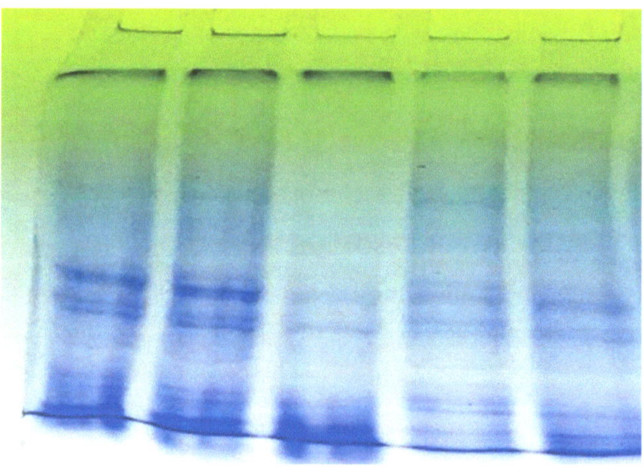

Fig: 4.1: 12% SDS-PAGE gel showing protein profiles of five varieties ofdate palm.(KHZ) Khenaizi, (SUG) Sugay,(MSH) Mishriq (KHD) Khadhrawi, and (BAH) Barhi.

Table: 4.1, showed the number of protein bands, intensity, and molecular weight of the date palm varieties revealed by SDS-PAGE analysis. Khenaizi profile is divided into six protein subunit, among them two bands have maximum intensity 12-35 kda, four bands have medium, the molecularweight in range of 12-35 kda. In lane two, Sugay profile, is divided into seven protein subunit, among them two bands have maximum intensity 21- 35 kda, five bands have medium, the molecular weight in range of 12 -75kda. Khadhrawi profile is divided into seven protein subunit, all bands have medium intensity, and the molecular weight in range of 15-75 kda. While only three bands with medium intensity was observed in Mishriq profile with low molecular weight in range of 12-25 kda. Lastly Barhi with four protein subunit between the ranges of 15-75 kda and medium intensity of bands. (Table: 4.1, Fig: 4.1).

BANDS	LANE1	LANE2	LANE3	LANE4	LANE5	Mol.Wt(Kda)
	KHZ	SUG	MSH	KHD	BAH	
BAND 1	-	+	-	+	-	75
BAND 2		-	-	+		38
BAND 3	++	++		+	+	35
BAND 4	+	+	-----	+	-----	28
BAND 5	+	+	+	+	+	25
BAND 6	+	+	-	+	-----	18
BAND 7	+	+	+	+	+	15
BAND 8	------ ++	++	+	------	-----+	12
TOTAL	6	7	3-----	7	4	

Table: 4.1: Number of protein bands, intensity, and molecular weight of thedate palm varieties revealed by SDS-PAGE analysis (- absent --- low intensity + medium intensity + + high intensity).

In general the protein pattern of the studied varieties seemed slightly different; these differences were observed in the low protein content bands. the differences of protein bands intensity between varieties protein patterns should be taken in the consideration ,since there are differences among studied varieties in the intensity of protein bands ,which appeared higher in Khenaizi and Sugay than others varieties, lower in Mishriq and Barhi varieties. This agree with the conclusion, that the diversity of protein bands between varieties within species are generally low (Vaz et al, 2004; Hastuti and Prabang, 2009, Khalifa et al; 2012). Extraction protein from date palm tissue was very difficult due to toughness and the present of abundant secondary metabolites, there for the protein concentration of extract going tobe low and its results sometimes irreproducible (Mrabet et al., 2016).

In this study, 5 proteins exhibited six polymorphic bands, e.g., Maximum six proteins subunits were observed in protein profiles of Sugay, Khadhrawi and Khenaizi. Sugay and Khadhrawi, showing the maximum protein bands compared to other plant materials. While Barhi shows 3 proteins subunits. Among these bands 15, 25 kda protein subunits are common in all varieties, A major reason may be advanced for this polymorphism is the date palm genome being expected to contain multiple copies of many genes (Xuet al., 2006), The distinct biophysical properties that govern their migration behavior in the polyacrylamide gel therefore might be due to differences in the amino acid sequences. It is also possible that one gene product may undergo different co-and/or posttranslational modifications that affect protein mass and charge (Sarnighausen et al., 2004).

Electrophoresis protein pattern technique using date palm leaflets was wide used as a source of whole proteins for studies related to phylogenetic studiesor varieties identification (Ahmed and ALqaradawi,2009: Dakhil et al:2013,AL-Issa,2013,Khaierallah et al: 2014).

The protein pattern is biochemical markers, and this marker represents biochemical traits that could be analyzed by protein electrophoresis. The differences in the amino acid composition of soluble proteins were used to investigate the genetic variation within species and phylogenetic relationships between species. Buvanendran V, Finney DJ. (1967).

The application of this marker was limited because the proteins are not genetic materials. They are products of gene expression, so they could be affected by environmental factors Drinkwater RD, Hetzel DJS. (1991). Thus,the direction attention of researchers is converted to the molecular markers. The molecular markers are based on the nucleotide sequence mutationswithin the individual"s genome; they are the most reliable markers available Yang, W et al. (2013).

4.2 ISSR-markers analysis

ISSR-PCR markers are efficient tools for varieties identification and estimation of relatedness through DNA fingerprinting; many ISSR primers were tested in the literature and applied on date palm genotypes (Zehdi et al;2002; Mitra et al; 2011).

4.3 Qualitative and quantitative analysis of genomic DNA

The quality and quantities of DNA may be important for the success of highly sensitive ISSR application .In this work a manual protocol CTAB for DNA extraction was successfully adapted, the scale and purity of DNA extracted by this protocol was satisfactory concentration 25-122 ng/ul, no degradation of DNA was observed, the quality of extracted DNA was (260/280 OD ratio = 1.5. Quality of DNA was further confirmed on a garose gel electrophoresis Fig. 4.2.

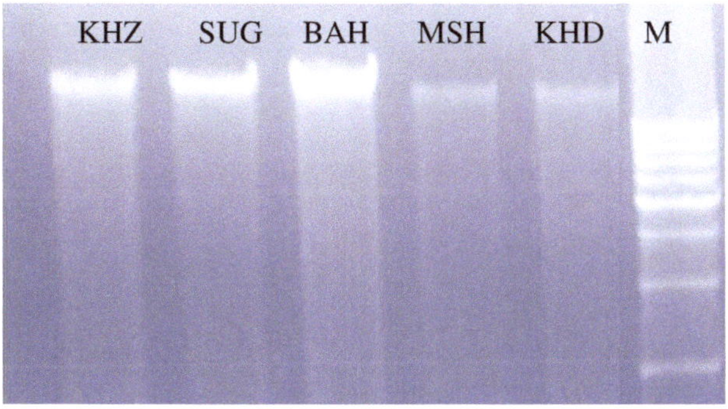

Fig. 4.2:0.8 % Agarose gel electrophoresis of total DNA extracted.

M: 100bp ladder marker. (KHZ) Khenaizi, (SUG) Sugay (BAH) Barhi (MSH) Mishriq and (KHD) Khadhrawi.

4.4 Inter Simple Sequence Repeat (ISSR) markers analysis The ISSR-PCRanalysiswas carried out in order to identify DNA polymorphism. Various parameters including No. of monomorphic and polymorphic bands, Polymorphism% etc., were calculated which areefficient for polymorphism analysis.

4.5 Individual primer results of date palm varieties

All primers have been discussed in separate sub heads for convenience according to the series and followed by overall results of the analysis.

(1) Primer DPISR1: the Non –anchored (CAG) tri- nucleotides Table 4.5.1,

4.5.2 and Fig 4:3 A) was annealed with 6 loci across all varieties, among them one was polymorphic, this primer produced total 29 bands the highest efficiency (18.4 %,) with 16.6 % polymorphism

The molecular weight of fragments or bands ranged from 380 bp to 900 bp and all were monomorphic except fragment 900 bp. On the other hand polymorphic band with discrimination power (2%) indicated the rareness of such repeat among the five varieties analyzed. The total number of amplifiedfragment for Khenaizi, Sugay, Khadhrawi, and Mishriq were 6, while Barhi5 bands.

bands	KHN	SUG	KHD	MSH	BAH
900	1	1	1	1	0
720	1	1	1	1	1
560	1	1	1	1	1
520	1	1	1	1	1
420	1	1	1	1	1
380	1	1	1	1	1
total	6	6	6	6	5

Table – 4.5.1: ISSR profile obtained by primer DPISR1:

No. of loci	6
Polymorphic fragment (no)	1
Monomorphic fragment (no)	5
Total bands(no)	29
% Polymorphism	16.6
% Discrimination power	2.0
% Primer Efficiency	18.4

Table –4.5.2: Parameters revealed by DPISR1 marker analysis

(2) Primer DPISR 2: The ISSR profile obtained with this primer (Table - 4.5.3, 4.5.4 and Fig 4:3 B) was annealed with 6 loci across all varieties and produced total 16 bands. The band with 200 bp was common for all varieties, the molecular weight of fragments or bands ranged from 200 bp to 700 bp. Two unique bands (520,490) bp were scored for Mishriq and Barhi respectively. The high polymorphism 83.3% value indicated a good discriminating capacity of the primer. The total number of bands for each variety range between 2 and 5.

bands	KHN	SUG	KHD	MSH	BAH
700	0	0	1	1	0
520	0	0	0	1	0
490	0	0	0	0	1
480	1	1	1	1	0
410	0	0	1	1	1
200	1	1	1	1	1
Total	2	2	4	5	3

Table - 4.5.3: ISSR profile obtained by primer DPISR2:

No. of loci	6
Polymorphic fragment (no)	5
Monomorphic fragment (no)	1
Total bands(no)	16
% Polymor-phism	83.3 %
% Discrimi-nation power	9.8
% Primer Efficiency	10.1

Table - 4.5.4: Parameters revealed by DPISR2 marker analysis

(3) Primer DPISR 3: The ISSR profile obtained this primer (Table – 4.5.5,

4.5.6 Fig 4:3 C) produced total 19 bands with 6 loci across all varieties with

83.3 % polymorphism and 9.8% Discrimination power. The band with 500 bp was common for all varieties; the molecular weight of fragments or bandsranged from 200 bp to 800 bp. one unique band (200) bp was scored for Mishriq. Thus, a genetic fingerprinting of this variety can be counted from this primer. The high polymorphism 83.3% value indicated a good discriminating capacity of the primer. The total number of bands for each variety range between 1 and 5.

bands	KHN	SUG	KHD	MSH	BAH
800	1	0	1	0	0
670	1	0	1	0	1
500	1	1	1	1	1
400	1	0	1	1	1
300	1	0	1	1	1
200	0	0	0	1	0
Total	5	1	5	4	4

Table – 4.5.5: ISSR profile obtained by primer DPISR3:

No. of loci	6
Polymorphic fragment (no)	5
Monomorphic fragment (no)	1
Total bands(no)	19
% Polymor-phism	83.3%
% Discrimi-nation power	9.8
% Primer Efficiency	11.4

Table –4.5.6 Parameters revealed by DPISR3 marker analysis

(4) Primer DPISR 4: The ISSR profile obtained (Table –4.5.7, 4.5.8 and Fig 4:3 D) was annealed with 10 loci across 9 loci were polymorphic with90 % polymorphism all varieties and produced total 24 bands. The Two unique bands (200,210) bp were scored for Khenaizi variety (Table - 3.9).Thus, a genetic fingerprinting of this variety can be counted from this primer. The high number of polymorphic band with 17.6 % discriminating power indicated the efficiency of the primer. The total number of bands for each variety range between 2 and 7.

bands	KHN	SUG	KHD	MSH	BAH
600	1	1	1	0	0
520	0	1	1	0	0
480	1	1	1	0	0
430	0	0	1	0	0
410	1	1	1	0	0
400	0	0	0	1	1
300	1	1	1	1	1
230	1	1	1	0	0
210	1	0	0	0	0
200	1	0	0	0	0
total	7	6	7	2	2

Table –4.5.7: ISSR profile obtained by primer DPISR4:

No. of loci	10
Polymorphic fragment (no)	9
Monomorphic fragment (no)	1
Total bands(no)	24
% Polymor-phism	90%
% Discrimi-nation power	17.6
% Primer Efficiency	15.2

Table –4.5.8: Parameters revealed by DPISR4 marker analysis

(5) **Primer DPISR 5:** The ISSR profile obtained with primer DPISR5 (Table –4.5.9, 4.5.10 and Fig 4:3 E) was annealed with 12 loci across all varieties which was the highest number of location among the primers, with 100 % polymorphism, no monomorphic loci was scored .and produced total 26 bands., the molecular weight of fragments or bands ranged from 200 bpto 580 bp. Some bands were unique with some varieties such as two unique bands (580,370) bp were scored for Sugay variety. Thus, a genetic fingerprinting of this variety can be counted from this primer. The high number of polymorphic band with 23.5 % discriminating power indicatedthe efficiency of the primer. The total number of bands for each variety range between 4 and 7.

bands	KHN	SUG	KHD	MSH	BAH
580	0	1	0	0	0
560	0	0	0	1	0
500	0	0	1	0	0
490	0	0	1	0	1
410	1	1	0	0	0
400	1	1	0	0	0
370	0	1	0	0	0
300	0	0	0	1	1
280	1	1	1	0	1
220	0	0	1	1	1
210	1	1	1	0	0
200	0	1	1	1	1
Total	4	7	6	4	5

Table –4.5.9: ISSR profile obtained by primer DPISR5:

No. of loci	12
Polymorphic fragment (no)	12
Monomorphic fragment (no)	0
Total bands(no)	26
% Polymor-phism	100%
% Discrimi-nation power	23.5
% Primer Efficiency	16.5

Table – 4.5.10: Parameters by DPISR5 marker analysis

(6) **DPISR6:** The banding pattern generated through this primer (Table–4.5.11, 4.5.12 and Fig 4:3F) showed 8 loci across all varieties with 100 % polymorphism, no monomorphic loci was scored .and produced total 16 bands., the molecular weight of fragments or bands ranged from 280 bp to 850 bp. two unique bands (280,300) bp were scored for Khenaizi variety (Table - 3.13).Thus, a genetic fingerprinting of this variety can be counted from this primer. Also one unique band (580) bp for Barhi variety was scored .The high number of polymorphic band with 15.7% discriminating power indicated the efficiency of this primer. The total number of bands for each variety range between 2 and 4.

bands	KHN	SUG	KHD	MSH	BAH
850	0	0	1	0	0
780	1	1	1	0	0
650	0	1	1	1	0
580	0	0	0	0	1
520	0	1	1	1	1
400	1	0	0	1	0
300	1	0	0	0	0
280	1	0	0	0	0
Total	4	3	4	3	2

Table 4.5.11: ISSR profile obtained by primer DPISR6

No. of loci	**8**
Polymorphic fragment (no)	8
Monomorphic fragment (no)	0
Total bands(no)	16
% Polymor-phism	100%
% Discrimi-nation power	15.7
% Primer Efficiency	10.1

Table –4.5.12 Parameters by DPISR6 marker analysis.

(7) **DPISR7:** The ISSR profile obtained with this primer (Table - 4.5.13,

4.5.14 and Fig 4:3 G) produced total 13 bands which was the lowest number of bands in all primers, with 80 % polymorphism. The molecular weight of fragments or bands ranged from 350 bp to 800 bp. Only one monomorphic band was observed at 580 bp, no unique band was scored. The low numberof loci and polymorphism 80 % than other markers indicate lowdiscriminating power of this primer. The total number of bands for each variety range between 2 and 3.

bands	**KHN**	**SUG**	**KHD**	**MSH**	**BAH**
800	1	1	0	1	0
650	1	1	0	0	0
580	1	1	1	1	1
450	0	0	1	0	0
350	0	0	1	0	1
Total	3	3	3	2	2

Table –4.5.13: ISSR profile obtained by primer DPISR7:

No. of loci	5
Polymorphic fragment (no)	4
Monomorphic fragment (no)	1
Total bands(no)	13
% Polymor-phism	80%
% Discrimi-nation power	7.8
% Primer Efficiency	8.2

Table –4.5.14: Parameters revealed by DPISR7 marker analysis

(8) **DPISR8:** The ISSR profile obtained with this primer (Table – 4.5.15,

4.5.16 and Fig 4:3 H) produced total 16 bands, no monomorphic band was observed thus 100 % polymorphism scored. The molecular weight of fragments or bands ranged from 200 bp to 950 bp. Only one unique bandwas observed at 260 bp for Barhi variety. The high number of polymorphic band with 13.7% discriminating power indicated the efficiency of this primer. The total number of bands for each variety range between 2 and 5.

bands	KHN	SUG	KHD	MSH	BAH
950	0	1	0	1	0
900	1	0	1	0	1
400	0	1	0	1	0
390	1	0	1	0	1
280	1	0	1	0	1
260	0	0	0	0	1
200	1	0	0	0	1
Total	4	2	3	2	5

Table –4.5.15: ISSR profile obtained by primer DPISR8.

No. of loci	7
Polymorphic fragment (no)	7
Monomorphic fragment (no)	0
Total bands(no)	16
% Polymor-phism	100%
% Discrimi-nation power	13.7
% Primer Efficiency	10.1

Table –4.5.16 Parameters revealed by DPISR8 marker analysis.

4.6 Bands analysis

The results obtained through eight ISSR primers (as listed in Table:

4.6 Fig 4.3) showed the eight ISSR-PCR primers which were used to amplify DNA segments from five date palm varieties (Khenaizi, Sugay, Khadhrawi, Mishriq and Barhi), were annealed with 60 loci across all variety genomes with an average of 7 loci per primer with range of 200 to 950 bp. Among those loci scored, 51 loci were polymorphic with (85%) polymorphism for at least one of the varieties with an average of 6 polymorphic bands per primer.

A total of 159 bands from all analysis with an average of 19.8fragments per primer, which were enough for the identification and evaluation of genetic diversity and designing the phylogenetic treefor these five different date palm varieties (table: 4.6 and Fig 4:3 Ato H). These results are in agreement with those of Zehdi et al. (2004), using ISSR on Tunisian date palm that generated 100 bands were identified at 14 microsatellite loci with an average of 7.14 alleles per locus. Adawy et al. (2002) using seven ISSR primers generated 53 fragments ranging from 298 to 1200 bp in size.

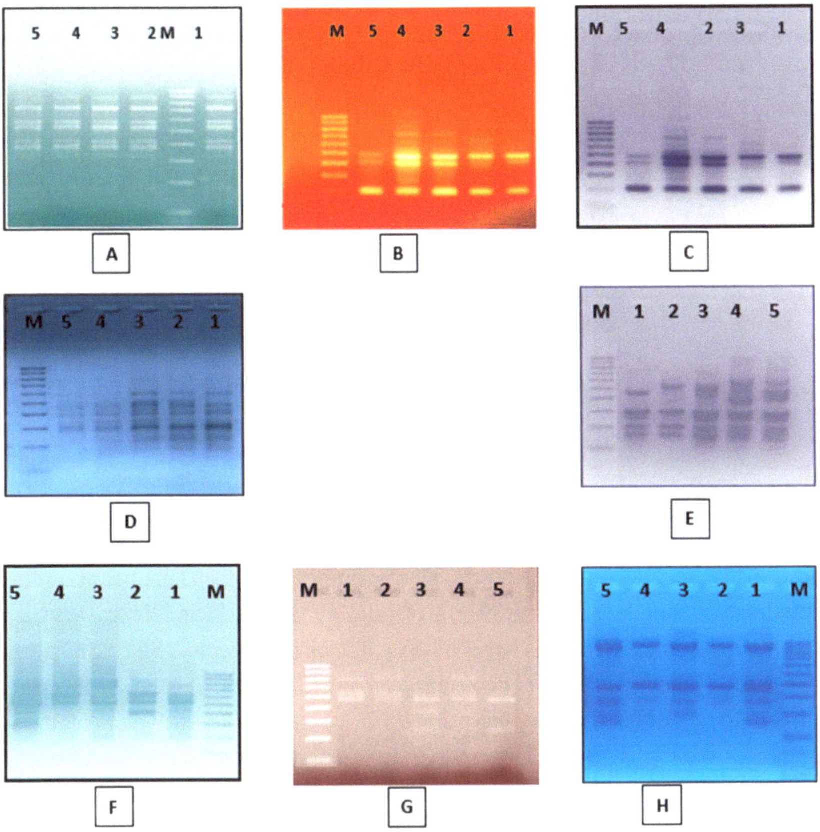

Fig. 4.3: ISSR profile of five date palm varieties amplified with eight different ISSR primers. (A) Primer DPISR-1, (B) Primer DPISR-2, (C)DPISR-3(D) DPISR-4,(E) DPISR-5 (F) DPISR-6, (G) DPISR-7and, (H)
DPISR-8.M: 100bp ladder marker. Lanes 1 through 5 refer to date palm varieties Khenaizi, Sugay, Khadhrawi, Mishriq and Barhi, respectively.

The average number of fragments per primer was 7.6 fragments with 64.1% polymorphism. While, Adawy et al. (2004) generated 159 fragments when using 19 ISSR primers to analyze bulked DNA samples representing five date palm varieties and the average number of fragments/primer were 8.4.

In Table: 4.6 the number of amplified fragments per primer varied from 13 bands for the primers DPISR7, which showed the lowest primers efficiency (8.2 %,) to 29 bands for the primer DPISR-1 which represented the highest efficiency (18.2 %,) Table 3.19 The most informative primers, considering percentage of polymorphism (%P=100) were DPISR-5, DPISR-6 and DPISR-8 with 12, 8 and 7 polymorphic bands respectively indicating their abundance over other in date palm genome while the non –anchored (CAG) tri- (DPISR1) nucleotides exhibited the lowest polymorphism (16.6%).

On the other hand polymorphic band with discrimination power (2%) indicated the rareness of such repeat among the five varieties analyzed. This is due to the absent of sites that complement the sequences of these primers in the palm genome and the extent of polymorphism varies with the nature of primer used and the sequence of repeats (motif) in the primer employed This result, together with those obtained by Ruas (2003), González et al. (2005) and Pérez de la Torre et al. (2010) indicate that the level of polymorphism detected by ISSR primers depends on the species or genusand the repetitive ISSR used in the primer utilized to generate the amplification profiles, additionally This result is in agreement with what wasreported by Zhao et al; 2012 who stated that the AG – ISSR repeat is the most abundant and polymorphic among di-nucleotide and comprises 85.7% of date palm genome.

Primer	No. of loci	NMB *	NPB *	TBN *	P%*	% E *	% D *	ABL(bp) *
DPISR1	6	5	1	29	16.6	18.2	2.0	380-900
DPISR2	6	1	5	16	83.3	10	9.8	200-700
DPISR3	6	1	5	19	83.3	12	9.8	200-800
DPISR4	10	1	9	24	90	15.1	17.6	200-600
DPISR5	12	0	12	26	100	16.3	23.5	200-580
DPISR6	8	0	8	16	100	10.1	15.7	280-850
DPISR7	5	1	4	13	80	8.2	7.8	350-800
DPISR8	7	0	7	16	100	10.1	13.7	200-950
Total	60	9	51	159	85	100	100	
Average	7	1.2	6	19.8	81.7	12.5	12.5	

Table 4.6: List of the used primers and the complementary information of the ISSR assay

*TBN: Total band number, NPB: Number of polymorphism bands, NMB: Number of monomorphism bands, P%: Polymorphism percentage, ABL: The amplified band length, D %: Discrimination power, E %: Primer Efficiency.

It is clear that the ISSR markers were differed among them in the number of bands according to the marker. This was shown by several studies, Karim, K el a;(2010), studied the genetic convergence between the ten date palm varieties in Tunisia to find the genetic convergence between the ten date palm varieties they used the same primers, only seven of them gave a result with 11 bands per primer (AG10C, AG10T, CT10A,CT10T, CT10A ,CT10G and GACA4).

4.7 Unique profiles

Some genotypes revealed a unique profile with the primers and the results were reproducible and this can be used for DNA fingerprinting and variety identification. Different primers produced unique bands for each variety in table: 4.6, DPISR-2 primer gave one unique band for each of Mishriq and

Barhi, at 520, 490 bp respectively. DPISR-3 gave one unique band forMishriq at 200 bp .DPISR-4 gave two unique bands for Khenaizi at 210, 200, bp, and one unique band for Khadrawi at 430 bp. DPISR-5, gave one unique band for each of Khadrawi and Mishriq at 500, 560 bp respectively, and two unique bands for Sugay at 580,370 bp. DPISR-6, gave one unique band for each of Khadrawi and Barhi at 850, 580 bp respectively, and two unique bands for Khenaizi at 300 280 bp. DPISR-7, gave one unique bands for Khadrawi at 450 bp, and DPISR-8 gave one unique band for Barhi at 260bp. The presence of unique ISSR markers among the various date palm varieties indicates the advantage of this approach for fingerprinting purposes. In this context, Adawy et al. (2002; 2004) revealed unique markers characterizing each of four varieties from Delta, Egypt and each of five varieties from Upper Egypt, respectively, the total number of unique ISSR markers was 24.

The appearance of unique bands may be explained on the basis of mutational events at the regulatory system of unexpected gene(s) or on the basis of bandsub fractionation which could be attributed to duplication of gene followed by the occurrence of point mutation that encoded the fractionated band. Wrang et: al (2002).

Abdullah and Gamal (2010) applied three types of markers such as protein RAPD and ISSR on four important varieties in Saudi Arabia, Inter varietal variation was investigated using five RAPD and five ISSR markers were also applied to assess the genetic polymorphism. Most of the varieties had the narrow genetic diversity as already expected.

Variety	Size of the markerband (bp)	Primer	Total markers / variety
Khenaizi	210, 200	DPISR- 4	2
	300, 280	DPISR- 6	
Sugay	580, 370	DPISR- 5	1
Khadrawi	430	DPISR- 4	4
	500	DPISR- 5	
	850	DPISR- 6	
	450	DPISR- 7	
Mishriq	200	DPISR- 3	3
	560	DPISR- 5	
	520	DPISR- 2	
Barhi	490	DPISR- 2	3
	580	DPISR- 6	
	260	DPISR-	

Table 4.7: The date palm varieties characterized by unique positive ISSR

markers, marker size and total number of markers identifying each variety.

The emergence of the unique bands is one of the most important diagnostic qualities adopted in isolating the species as the emergence of the bands appears in a particular type indicates the presence of that sequence in the DNA genome of that species without other species as well as the possibility of obtaining the genetic fingerprint of some of these types through the presence of this bands (Wrang et: al 2002).

4.8 Genetic distance and Phylogenetic tree analysis

In the present study ISSR, SDS-PAGE and combined analysis data was converted into a binary data in excel work sheet and was analyzed using SPSS-16 program to find the genetic distance between and within the five different date palm varieties. Table 4.8.1, 4.8.2 and 4.8.3three trees generated from the three analyses showed six similar clusters Fig. 4.8.1,4.8.2 and 4.8.3).

The genetic distances matrix was calculated for the 51, 8 polymorphic bandsfor ISSR, protein pattern respectively and combined analysis of the five varieties on the basis of present and absent of the polymorphic bands. The genetic distance and separation of each variety varied according to the typeof analysis used. The range is between (0.922-0.645) table 4.8.1, (0.816- 0.289) table 4.8.2 and (0.887-0.627). Table 4.8.3 with a mean of (0783- 0.552-0.757) for the three analyses respectively. Thus the accessions testedin the study are high divergent at the DNA level.

The smallest distance value was observed between Khenaizi and Sugay varieties (0.289, 0.627) for the SDS-PAGE and combined based analysis respectively, which appear to be the most similar varieties and can be closelyrelated. While Sugay variety was highly divergent from Barhi variety with distance (0.922 - 0.887) for the ISSR and combined based analysis respectively.

It is noteworthy that Mishriq presented very limited average distance (from 0.780 to 0.849) with Khenaizi, Sugay and Khadhrawi, thus Mishriq could be characterized by a high divergence at the DNA level and could be unlikely regrouped with them. All the varieties display different intermediate level of dissimilarity (0.627 to 0.658) and are grouped with the other ones.

	Matrix File Input				
	KHENAIZI	SAGAY	KHADHRAWI	MISHRIQ	BARHI
KHENAIZI	.000				
SAGAY	.664	.000			
KHADHRAWI	.645	.683	.000		
MISHRIQ	.852	.774	.751	.000	
BARHI	.797	.922	.659	.761	.000

Table 4.8.1 Genetic distance values of Nei's coefficient revealed byISSR analysis

	Matrix File Input				
	KHENAIZI	SAGAY	KHADHRAWI	MISHRIQ	BARHI
KHENAIZI	.000				
SAGAY	.289	.000			
KHADHRAWI	.516	.408	.000		
MISHRIQ	.548	.577	.816	.000	
BARHI	.447	.500	.707	.408	.000

Table 4.8.2 Genetic distance values of Nei's coefficient revealed by SDS-PAGE analysis

	Matrix File Input				
	KHENAIZI	SAGAY	KHADHRAWI	MISHRIQ	BARHI
KHENAIZI	.000				
SAGAY	.627	.000			
KHADHRAWI	.637	.658	.000		
MISHRIQ	.849	.780	.793	.000	
BARHI	.780	.887	.701	.749	.000

Table 4.8.3 Genetic distance values of Nei's coefficient using combined dataof

ISSR and SDS-PAGE analysis

The trees constructed in Fig 4.8.1, 4.8.2 and 4.8.3 explained the molecular phylogenetic relationships between the five varieties. Unweight Pair Group Method with Arithmetic Average (UPMGA) analysis ordered date palm varieties into two main clusters in ISSR, SDS-PAGE, and ISSR, SDS-PAGEcombined analyses.

According to combined analyses that revealed by ISSR, SDS-PAGE combined analyses.59 polymorphic loci, UPGMA ordered the date palm varieties into two main clusters, Fig 4.8.3 the first one consisted of two varieties: Mishriq and Barhi it is similar to those based on agronomic traits. And the second one consisted of two sub-clusters, the first one consisted of two varieties, Khenaizi and Sugay and the second sub-cluster consisted of Khadhrawi variety. Fig 4.8.3

UPGMA ordered the 51 polymorphic loci of ISSR analysis of the date palm varieties into two main clusters the first one contained two varieties: Mishriqand Barhi. While Khenaizi, Sugay and Khadhrawi grouped in the second onewhich consisted of two sub-clusters, the first one consisted of Khenaizi and Khadhrawi, the second sub-cluster consisted of Sugay variety Fig 4.8.1.

It is interesting to observe that combined tree of ISSR and SDS-PAGE analysis in Fig 4.8.3 ordered the date palm varieties under study into two main clusters according to UPGMA analysis as well was highly similar tothe SDS-PAGE analysis tree Fig 4.8.2. The first one consisted of two varieties: Mishriq and Barhi, which were closely related varieties. And the second one consisted of two sub-clusters, the first one consisted of two varieties, Khenaizi and Sugay and the second sub-cluster consisted of Khadhrawi variety.

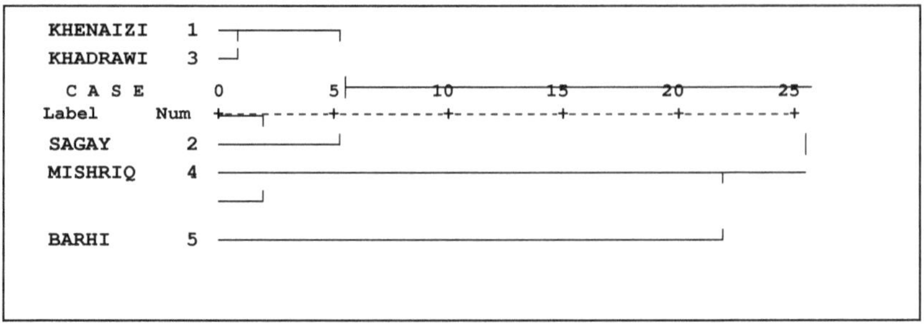

Fig 4.8.1 Cluster analysis with UPGMA method of related five study datepalm varieties using ISSR, data based on Jaccard similarity matrix.

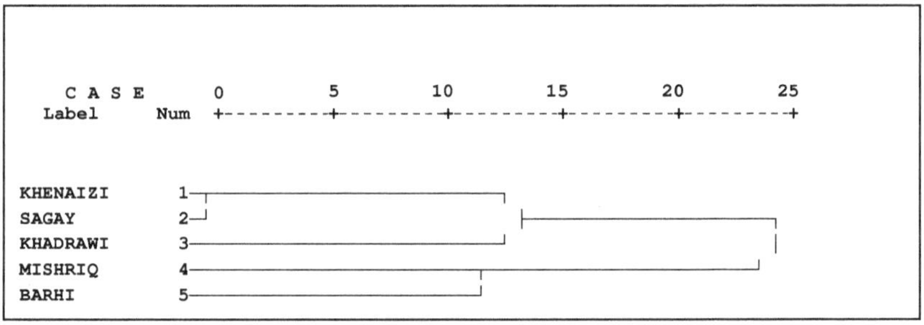

Fig 4.8.2 Cluster analysis with UPGMA method of related five study datepalm varieties using SDS-PAGE data based on Jaccard similarity matrix.

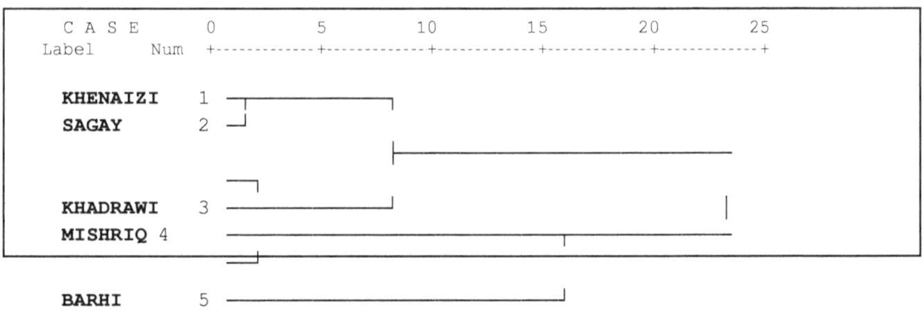

Fig 4.8.3. Cluster analysis with UPGMA method of related five study date palm varieties using combined data (ISSR, + SDS-PAGE) based on Jaccardsimilarity matrix.

Generally, the genotypes which tested revealed that the geographic origin was not affected the cluster divisions. Accordingly, the sister varieties `Khenaizi` and 'Sugay with (0.289, 0.627 distances Table 8.4.2 and 8.4.3) which have different geographical origins (The United Arab Emirates and the Kingdom of Saudi Arabia respectively) fell in one sub-cluster, Fig 8.4.3 and 8.4.2. Also the two varieties Mishriq and „Barhi" (Sudan and Iraq, respectively) fell in the same cluster, This result agrees with other reports for Moroccan, Algerian and Tunisian date palm varieties based on analyses using microsatellite markers (Zehdi et al., 2004) and isozyme markers (Ould Mohamed Salem et al., 2001).

A high degree of independence between the geographical origin and molecular data was indicated. The RAMPO and AFLP data applied onTunisian date palms (Rhouma et al., 2007; Rhouma-Chatti et al., 2011) showed that the studied varieties clustered independent on their geographic origin. The two varieties Mishriq and „Barhi" (Sudan and Iraq, respectively) which related to the same cluster, based on agronomic traits particularly the fruits which were characterized by dates of medium size and brown color. This could be explained by the presence of a common genetic origin among the tested varieties in spite of their origin and this agree with Hammadi et al. (2012) who found that fruit consistency which is an important characteristic of date fruit has association with genetic markers because clustering basedon fruit consistency is in accordance with clustering by microsatellite markers.

The varieties Barhi and 'Khadhrawi originated in Iraq clustered in a different cluster this observation suggested that genetic variation range with each of the Iraqi varieties exits grouped, this result confirmed what was obtained byAl-Najm et al; (2017) when they used inter-primer binding site (iPBS) markers to assess the molecular variation and genetic diversity of 54 and 12 date palm genotype collected from Australia and Iraq, they found that Barhi and 'Khadhrawi originated in Iraq in a different group to those collected in Iraq ,their observations suggests that a range of genetic variation within each of the Iraqi varieties exists. So this could be helpful In Sudan where date palm breeding is highly dependent on seed propagation with subsequentselection based on specific characteristics such as fruit quality and plant vigor as determined by local farmer preferences (Khierallah et al., 2011).

Results also showed that the foreign date palm varieties recently introduced to Sudan groves are closely grouped in one cluster this could be explainedby the presence of a common genetic origin among them, Al-Khalifah and Skari (2013), added that over the years many date palm varieties have been transplanted to areas other than the area of their origin, and there may have been adapted with different names. On the whole our data augment those describing the application of molecular tools in date palm variability analysis and previously reported (Trifi et al; 2000) Dendogram showed that accessions grouping in relation to their geographical origin are not well defined.

The results of the present study revealed that the average polymorphism detected by the ISSR assay (85 %) was higher than that observed for protein pattern (75%). This was also evidenced through the number of varieties that could be identified by each technique through unique fragments. ISSR data allowed the discrimination of five varieties. However, the use of SDS-PAGE, in spite of low numbers of bands could be distinguished Khadhrawi and Mishriq varieties by negative unique bands at 12,35 kda respectively (Table 3:1), while ISSR assay could be distinguished the five varieties through 16 unique bands. (Table 3:20), ISSR is a simplest, quickest marker system with high reproducibility, there for it could has an important role in distinguishing even closely related germplasm (Reddy et al; 2002).it is clear from the literature that ISSRprimers could detect more polymorphism than with mtDNA, cpDNA, RAPDand isozyme in closely related plants (Virk et al;2000).In the field of date palms, ISSR markers were found more informative than the RAPD markers (Mitra et al., 2011).

Many ISSR primers were tested in the literature and applied on date palm genotypes (Zehdi et al., 2002; Mitra et al., 2011) or on other monocotyledon species such as the genus Poa (Arslanet al., 2011) and durum wheat (Pasqualone et al., 2000). Thus ISSR profiles can be of much use in developing molecular IDs for each date palm varieties for its proper identification, registration and conservation.

CHAPTER FIVE CONCLUSIONS AND RECOMMENDATIONS

5.1 Conclusions

(1) Determination of protein molecular weight via polyacrylamide gel electrophoresis (PAGE) in the present of sodium dodecyl sulfate (SDS) whatever is a classic study to differentiate between varieties, is a universal can be economically used to assess genetic variation and relation in germplasm.

(2) Proteins are primary gene products of active structural genes; their size and amino acids sequence are the direct results of nucleotide sequences ofthe genes; hence, any observed variation in protein systems induced by any mutagen is considered a mirror for genetic variations.

(3) Our data provide from this study evidence of a genetic diversity between the tested varieties indicating the interrelationship between the date palm varieties.

(6) Some genotypes revealed a unique profile with the primers and the results were reproducible and this can be used for DNA fingerprinting and variety identification.

(7) The date palm varieties recently introduced to Sudan have a common genetic origin, they closely grouped in one cluster.

(8) The most informative primers, considering percentage of polymorphism (%P=100) were DPISR-5, DPISR-6 and DPISR-8 with 12, 8 and 7 polymorphic bands respectively indicating their abundance over other in date palm genome while the Non –anchored (CAG) tri- (DPISR1)nucleotides exhibited the lowest polymorphism (16.6%).

(9) The ISSR marker revealed a higher ratio of 85% polymorphism than protein. The variation obtained by these types of markers might be because different marker systems differ in the mechanism of detecting polymorphism, genome coverage and the ease of application.

(10) We suggest that these varieties are not a result of intact cloning processes. We expect that the propagation cycles performed by farmers weredisrupted by a mixture of plants derived from seeds.

(11) Hand pollination is mainly mechanical; the farmers depend on few selected males for the pollination of female trees. Male trees are selected for fruit quality and are exchange between farmers (Osman, 1984).

Pollen grains of more than one male are sometimes mixed and used for pollination. New varieties are the result of a continuous selection processthat is carried out by farmers in their fields following sexual reproduction. The exchange of propagules, which are a mix of vegetative and seed- propagated materials, is conducted between farmers. All these processes together may result in a mixed date palm genome within the country.

5.2 Future recommendations

(1) Our data indicate that the use of molecular tools, such as SDS-PAGE and ISSR are very important to explain genetic diversity and population structures. In the future, the use of these tools should increase in order to fully investigate the relationships between date palm varieties.

(2) Date palm in Sudan needs a sustainable management to contribute to the conservation of Sudanese date palm where the biodiversity is now endangered by biotic and abiotic stresses.

(3) This is the first study to address genetic diversity of five date palm varieties from Sudan using the SDS-PAGE and the eight ISSR markers more researches should do for date palm improvement through biotechnological research to identify and quantify genetic diversity components in the species,identify and clone genes and gene complexes for biotic and abiotic stresses and utilize the generated information for future research and development.

(4) Developments in genomics have provided new tools for discovering and tagging novel alleles and genes. These tools can enhance the efficiency of breeding programs through their use in marker-assisted selection.

(5) Genetic diversity is desirable for long-term crop improvement and reduced vulnerability to important crop diseases. Measurements of genetic diversity can be used in breeding programs to increase the genetic variation in populations by crossing varieties with a high level of genetic distance and for the introgression of exotic germplasm.

(6) In the future Gene banks should be set up for the effective conservation of date palm varieties because crop information is very important in modern plant breeding, particularly when advanced research institutions, and private sector breeding companies share in this case.

(7) The palm date diversity, maintained by generations of growers within thesort of in place or on-farm conservation using lore , is of no use nowadays without the right interest of growers and encouragement of state organizations and hence urging the importance of adoption of recent way forconservation.

REFERANCES

Abdulla M. and Gamal O. (2010).Investigation on molecular phylogeny of some date palm (Phoenix *dactylifera* L.) varietys by protein, RAPD and ISSR markers in Saudi Arabia. Aust. J Crop Sci. 4: 23–28.

Aebersold R, Goodlett DR. 2001. Mass spectrometry in proteomics. Chem Rev 101: 269– 295.

Adawy, S.S., Hussein, Ebtissam H.A., ElKhishin, Dina, Moharam, H. and ElItriby, Hanayia A. (2002). Genetic variability studies and molecular fingerprinting of some Egyptian date palm (Phoenix dactylifera L.) varietys: II. RAPD and ISSR profiling. Arab J. Biotech., 5(2): 225-236.

Adawy, S.S., Hussein, Ebtissam H.A., Saker, M.M. and El-Itriby, Hanaiya A. (2004). Intra- and Inter-varietal variation of Upper Egypt date palmvarietys (Phoenix dactylifera L.): I. As revealed by RAPD and ISSR markers. Proceed. Int. Conf. Genet. Eng. & Appl., Sharm El-Sheikh, South Sinai, Egypt (April, 8-11, 2004): 165-179.

Ahmed TA, Al-Hadidi S. (2014). Molecular characterization of date palm (Phoenix dactylifera L.) using Inter Simple Sequence Repeat (ISSR) markers. In:Zaid A, Alhadrami GA. (eds). Proceedings of the Fifth International Date Palm Conference. Abu Dhabi, UAE. March 16-18. Khalifa International Date Palm Award. pp: 161.

Ahmed, S. D., Sabir, S. M., Saud, H. M., and Salihuddin, Y. (2008): Evolutionary Relationship and Divergence Based on SDS-PAGE, of Elaeagnus umbellata (Thunb.) Populations, a Multipurpose Plant from the Himalayas.Turk. J. Biol. 32, 31-35.

Ahmed TA, Al-Qaradawi A. 2009. Molecular phylogeny of Qatari date palm genotypes using simple sequence repeats markers. Biotechnology 8: 126-131.

Akkak A, Scariot V, Torello Marinoni D, Boccacci P, Beltramo C, Botta R. (2009). Development and evaluation of microsatellite markers in Phoenix dactylifcra L. and their transferability to other Phoenix species. Biol Plant 53(1): 164-166.

Al-Dous, EK, George B, Al-Mahmoud ME, Al-Jaber MY, Wang H,Salameh YM, Al-Azwani EK, Chaluvadi S, Pontaroli AC, DeBarry J,Arondel V, Ohlrogge J, Saie IJ, Elmeer KMS, Bennetzen JL, Krueger RRand Malek JA.

(2011). De novo genome sequencing and comparativegenomics of date palm (Phoenix dactylifera). Nat. Biotechnol. 29: 521-527. **AL-Ekidy**, H. (2000). The Date Palm, Amman, Hashemite Kingdom ofJordan: Zahran Press. pp. 340-389.

Al-Faifi SA, Migdadi HM, Algamdi SS, Khan MA, Ammar MH, Al-Obeed RS, Al-Thamra MI, El-Harty EH, Jakse J. (2016). Development, characterization and use of genomic SSR markers for assessment of genetic diversity in some Saudi date palm (Phoenix dactylifera L.) varietys. Electronic Journal of Biotechnology 21: 18-25.

Al-Farsi, M., Alasalvar, C., Al-Abid, M., Al-Shoaily, K., Al-Amry, M. & Al-Rawahy, F. (2007). Compositional and functional characteristics of dates,syrups, by-products. Food Chemistry, 104, 943–947.

Al-Harrasi, A. Rehman, N. U. Hussain, J. Khan, A. L., Al-Rawahi, A., Gilani, S. A. et al. (2014).Nutritional assessment and antioxidant analysis of 22 datepalm (Phoenix dactylifera) varieties growing in Sultanate of Oman. Asian Pac.J.Trop.Med. 7, S591–S598.doi: 10.1016/s1995-7645(14)60294-7.

Al-Helal, A. A. (1994): Electrophoretic analysis of protein and selected enzymes of Acacia from Saudi Arabia JKAU (Sci.) 6:17-28.

Al-Khalifah, N. S., Askari, E., & Shanavaskhan, A. (2013). Date palmtissue culture and genetical identification of varietys grown in Saudi Arabia. National Center for Agriculture Technologies, King Abdulaziz City for Science and Technology, Kingdom of Saudi Arabia.

Al-Khalifah NS (2006). Metaxenia: influence of pollen on the maternal tissue of fruits of two varietys of date palm (Phoenix dactylifera L.). Bangladesh J Bot 35: 151-161.

Al-Khayri, Jameel M Abdulaziz M Al-Bahrany 2004/2/6 Scientia Horticulturae V,99 Issue 2. Page 153-162. Pub.Elsevier .

AL-Issa AM (2013) Electrophoretic analysis of protein pattern in date palm "Khalas" variety leaflets among different locations of AL-Ahsa .Am .J Agric.Bio.Sci.8(4):343-349.

Al-Mssallem, I.S., Hu, S., Zhang ,X., Lin, Q., Liu, W., Tan, J., et al. (2013). GenomesequenceofthedatepalmPhoenixdactyliferaL.Nat.Commun.4:2274.d oi: 10.1038/ncomms3274.

Al-Najm A, Luo S, Ahmad NM, Pourkheirandish M, Trethowan R. (2017) . Molecular variability and population structure of a core collection of date palm (Phoenix dactylifera L.) varietys from Australia and the Middle East. AJCS 11(09):1106-1115.

Al-Qarawi, A. A., Mousa, H. M., Ali, B. H., Abdel-Rahman, H., and El- Mougy,

S. A. (2004). Protective effect of extracts from dates (Phoenixdactylifera L.) on carbon tetrachloride–induced hepatotoxicity in rats. Inter.
J. Appl. Res. Vet. Med. 2, 176–180.

Al-Shahib, W., and Marshall, R. J. (2013). The fruit of the date palm: it"s possible use as the best food for the future? Int. J. Food Sci. Nutr. 54, 247– 259. doi: 10.1080/09637480120091982.

Al-Yahyai, R. and S. Al-Khanjari. 2008. Biodiversity of date palm in the Sultanate of Oman. Afr. J. Agric. Res. 3(6):389–395.

Arslan E, TamkocA (2011)The application of ISSR-PCR to determine the genetic relationship and genetic diversity between narrow leaved bluegrass (Poaangustifolia) and rough bluegrass (Poatrivialis) accessions.Turk J Biol 35: 415–423.

Baaziz, M. & M. Saaidi. (1988). Preliminary identification of date palm varietys by esterase isoenzymes and peroxidases activities. Canadian Journalof Botany 66: 89-93.

Baaziz M, Majourhat K, Bendiab K (2000) Date palm culture in the Maghreb countries: constraints and scientific research. Proceedings of the date palm international symposium, Windhoek, Namibia, 22–25 Feb, pp 306–311.

Barghini, P., Di Gioia, D., Fava, F., and Ruzzi, M. (2007). Vanillin production using metabolically engineered Escherichia coli under non- growing conditions. Microb.Cell Fact. 6, 1475–2859. doi: 10.1186/1475- 2859-6-13.

Billotte N, Marseilla N, Brottier P, Noyer JL, Jacquemoud-Collet JP, Moreau C, Couvreur T, Chevallier MH, Pintaud JC, Risterucci AM. (2004). Nuclear microsatellite markers for the date palm (Phoenix dactylifera L.): characterization, utility across the genus Phoenix and in other palm genera. Mol Ecol Notes 4(2): 256-258.

Bodian A, Elhoumaizi MA, Nadir KN, Hasnoui A, Nachtigall M, Wehling P. (2012a). Genetic diversity analysis of date palm (Phoenix dactylifera L.) varietys from Figuig oasis (Morocco) using SSR markers. IJSAT 2: 96-104.

Bodian A ,NdoyeNdir K , Diedhiou PM, Borgel A,Sane D,Sagna M (2012b) Analysis of genetic diversity of date palm (Phoenix dactylifera L) varietys from Mauritania using microsatellite markers.Int J Sci Adv Technol 2:36-43.

Botstein, D.,R. L.White, M. Skolnickand R. W. Davis. (1980). Construction of a genetic linkage map in man using restriction fragment length polymorphisms. Am J Hum Genet 32: 314-331.

Buvanendran V, Finney DJ.(1967). Linkage relationships of egg albumen loci in the domestic fowl. Br Poult Sci., 8:9–13.

Castillejo MA, Curto M, Fondevilla S, Rubiales D, Jorrin JV. (2012). Two-dimensional electrophoresis based proteomic analysis of the pea (Pisumsativum) in response to Mycosphaerella pinodes. Journal of Agricultural and Food Chemistry 58: 12822–12832.

Chao, C. T., & Krueger, R. R. (2007). The date palm (Phoenix dactylifera L.): overview of biology, uses, and cultivation. HortScience, 42(5), 1077- 1082.

Chang, I.F., P.J. Chen, C.H. Shen, T.J. Hsieh and Y.W. Hsu et al., (2010). Proteomic profiling of proteins associated with the rejuvenation of Sequoia sempervirens (D. Don) Endl. Proteome Sci., 8: 64- 64. PMID: 21143964.

Chemelík, J. A. Krumlová, M. Budinská, T. Kruml, V. Psota, I. Bohaèenko, P. Mazal and H. Vydrová. 2002. Comparison of size characterization of barley starch granules determined by electron and optical microscopy, low laser light scattering and gravitational field-flow fractionation. J. Inst. Brew. 107:11–17.

Comas CI, Hunziker JH, Crisci JV.(1979). Species relationships in Bulnesiaas shown by seed protein electrophoresis. Biochem Syst Ecol 7 (4): 303-308.

Cordwell S. J., Nouwens A. S., Walsh B. J. 2001; Comparative proteomics of bacterial pathogens. *Proteomics*1:461–472.

Dakhil S, Coquel, Cosette P, ELKahoui S,Song PTC,Vaudry D, Hannachi AS, Trifi N , Limam F, Jouenne T,(2013). The date Palm (Phoenix dacylifera L) leaf proteoms .Identification of gender biomark to screen male parents Plant Omics J 6 (1) 18-23.

Davar, R., R. Darvishzadeh and A. Majid, (2012).Sclerotinia-induced accumulation of protein in the basal stem of resistant and susceptible lines of sunflower. Not. Bot. Horti. Agrobo., 40: 119-124.

Diaz S, Pire C, Ferrer J, Bonete MJ. 2003. Identification of Phoenix dactylifera L. varieties based on amplified fragment length polymorphism (AFLP) markers. Cell Mol Biol Letters 8: 891-900.

Drinkwater ,R.D.and Hetzel.D.J.S(1991).Application of molecular biology to understanding genotype-environment interactions in livestock production .In:Proc.of an international Symposium on Nuclear Techniques in Animal Production and Health IAEA,FAO.Vienna,15-19 April 1991:437-452.

Ehsanpour AA, Shojaie B, Rostami F (2010). Characterization of seed storage protein patterns of four Iranian Pistachios using SDS-PAGE. Nat. Sci. 2(7): 737-740.

El-Assar AM, Krueger RR, Devanad PS, Chao CT (2005) Genetic analysis of Egyptian date (Phoenix dactylifera L.) accessions using AFLP markers. Genet Res Crop Evol 52:601–607.

El-Hady, E.A.A.A., A.A.A. Haiba, N.R. Abd, A.A. Rizkalla and Elham et al., 2010.Phylogenetic Diversity and Relationships of Some Tomato Varieties by Electrophoretic Protein and RAPD analysis. J. Am. Sci., 6: 434-441.

Elhoumaizi MA, Saaidi M, Oihabi A, Cilas C (2002) Phenotypic diversity of date-palm varietys (Phoenix dactylifera L.) from Morocco. Genet Res Crop Evol 49:483–490.

Elhoumaizi MA, Devanand PS, Fang J, Chao C-CT. 2006. Confirmation of `Medjool' date as a landrace variety through genetic analysis of `medjool' accessions in Morocco. J Am Soc Hortic Sci 131: 403-407.

Elmeer K, Sarwath H, Malek J, Baum M, Hamwieh A. (2011). New microsatellite markers for assessment of genetic diversity in date palm (Phoenix dactylifera L.). 3Biotech. 1: 91-97.

Elshibli S, Korpelainen H (2008) Microsatellite markers reveal high genetic diversity in date palm (Phoenix dactylifera L.) germplasm from Sudan.Genetica 134:251–260.

Elshibli.S, and H. Korpelainen, "Biodiversity of date palms (Phoenix dactylifera L.) In Sudan: chemical, morphological and DNA polymorphismsof selected varietys," Plant Genetic Resources: Characterisation andUtilisation, vol. 7, no. 2, pp. 194–203, 2009.

Elshibli.S, and H. Korpelainen, (2009b) "Biodiversity of date palms (Phoenix dactylifera L.) In Sudan: chemical, morphological and DNA polymorphisms of selected varietys," Plant Genetic Resources: Characterisation and Utilisation, vol. 7, no. 2, pp. 194–203.

Emoghene B, Asemota O, Idu M, Eke C (2015) Molecular characterization of some date palms in Nigeria using RAPD markers. J App Biol Biotech 3:26–30.

Eoin, L.N.(2016). Systematics: blind dating. Nat. Plants 2:16069. Doi: 10.1038/nplants.2016.69.

Faqir N, Muhammad A, Shehzad A, Rahman H, Hyder MZ, Ali GM.(2016). Simple Sequence Repeat (SSR) markers show greater similarity among morphologically diverse Date palm (Phoenix dactylifera L.) varietys grown in Pakistan. Pure and Applied Biology 5: 483-498.

Filho JAF, de Brito LS, Leão AP, Alves AA, Formighieri EF, Júnior MTS. (2017).In silico approach for characterization and comparison of repeats in the genomes of oil and date palms. Bioinform Biol Insights11:1-12.

Garcia AAF, Benchimol LL, Barbosa AMM, Geraldi IO, Souza Jr. CL and Souza AP (2004) Comparison of RAPD, RFLP, AFLP and SSR markers for diversity studies in tropical maize inbred lines.Genetics and Molecular Biology.

Godwin, ID, Aitken, EAB & Smith, LW 1997, „Application of inter simple sequence repeat (ISSR) markers to plant genetics", Electrophoresis, vol. 18, pp. 1524–1528.

Gonzalez, A.; Wong, A.; Delgado- Salinas, A.; Papa, R. and Gepts, P. (2005). Assessment of inter simple sequence repeat markers to differentiate sympatric wild and domesticated populations of common bean. Crop Science, vol. 45, no. 2, p. 606-615.

Gorji AM, Poczai P, Polgar Z, Taller J.(2011). Efficiency of arbitrarily amplified dominant markers (SCoT, ISSR and RAPD) for diagnostic fingerprinting in tetraploid potato.Am. J. Potato Res. 2011; 88:226-237.

Gupta, M., Y-S.Chyi, J. Romero-Severson and J. L. Owen (1994) Amplification of DNA markers from evolutionarily diverse genomes using single primers of simple-sequence repeats.Theor. Appl. Genet. 89, 998- 1006.

Gupta PK, Varshney RK (2000) the development and use of microsatellite markers for genetic analysis and plant breeding with emphasis on bread wheat. Euphytica 113:163–185.

Haider, N., I. Nabulsi and M.N. Ali, 2012. Phylogenetic Relationshipsamong date palm (Phoenix dactylifera L.) varietys in Syria using RAPD and ISSR markers. J. Plant Biol. Res., 1: 12-24.

Hammadi H, Benabderrahim MA, Elbekkay M, Ferdaous G. Triki T, Ferchichi A. 2012. Investigation of genetic variation in Tunisian date palm(Phoenix dactylifera L.) varietys using ISSR marker systems and their relation with fruit characteristics. Turk J Biol 36: 449-458.

Hamwieh A, Farah J, Moussally S, Al-Sham'aa K, Almer K, Khierallah H, Udupa S, Lababidi S, Malek JA, Aaouine M, Baum M. (2010). Developmentof 1000 microsatellite markers across the date palm (Phoenix dactylifera l.)Genome. Acta Hortic 882: 269-277.

Hassan, S., K. Bakhsh, Z.A. Gill, A. Maqbool and W. Ahmad.(2006).Economics of growing date palm in Punjab, Pakistan. International Journal of Agriculture & Biology, 8 (6): 788–792.

Hastuti DS, Prabang S(2009) .Vairation morphology and protein band pattern of adenium (Adenium obesum) varieties Bioscience 1(2):78-83.

Harushima Y, Yano M, Shomura A, Sato M, Shimano T, Kuboki Y.(1998). A high-density rice genetic linkage map with 2275 markers usinga single F2 population.Genetics.1998; 148:479-494.

Hazzouri, K. M. et al. Whole genome re-sequencing of date palms yields insights into diversification of a fruit tree crop. Nat. Commun. 6, 8824 (2015).

Jaradat A.A (2015) Biodiversity of date palm. Land use, land cover and soil sciences. Encyclopedia Of life support systems (EOLSS), developed under the Auspices of the UNESCO, Eolss Publishers, Oxford, p 31.

Joshi SP, Gupta VS, Aggarwal RK, Ranjekar PK, Brar DS.(2000). Genetic diversity and phylogenetic relationship as revealed by inter simple sequence repeat (ISSR) polymorphism in the genus Oryza Theory Appl. Genet. 2000; 100:1311-1320.

Jubrael JMS, Udupa SM, Baum M.(2005). Assessment of AFLP based genetic relationships among date palm (Phoenix dactylifera L.) varieties of Iraq. J Am Soc Hortic Sci 130: 442447.

Karim. K, Chokri. B, Amel. H, Wafa. H, Richid. H and Nouredine, D. 2010. Genetic diversity of Tunisian date palm germplasm using ISSDR markers. Int. J. Bot. 6(2): 182- 186.

Khalifah NS, Ejaz A, Shanavas KAE (2012) Molecular and Morphological identification of some elite varities of date palm grown in Saudia Arabia ,Emir .J Food gric.24(5):456-461.

Khan A, Khan IA, Heinze B, Azim MK (2011) The chloroplast genome sequence of date palm (Phoenix dactylifera L. cv. „ Aseel „). Plant Mol Biol Rep 30:666–678.

Kheirallah H, Al-Sammarraie SKI, Haider IM (2014). Molecular characterization of some Iraqi date palms varietys using RAPD and ISSR markers, J. Asian Sci. Res. 4:490-503.

Khierallah H, Bader S, Baum M, Hamwieh A. 2011. Assessment of genetic diversity for some Iraqi date palms (Phoenix dactylifera L.) using amplified fragment length polymorphisms (AFLP) markers. Afr J Biotechnol 10: 9570-9576.

Khushk, A.M., A. Memon and K.M. Aujla. 2009. Marketing channels and margins of dates in Sindh, Pakistan. J. Agri. Res. 47:293-308.

Ladizinsky, G.Hymowitz, T.1979 Seed protein electrophoresis in taxonomicand evolutionary studies Theoretical and Applied Genetics 145-54-4.

Laemmli,U.K.(1970). Cleavage of structural proteins during the assembly of the heat of bacteriophage T4. Nature, 227: 680-685.

Leonardo, L.C.D., T.S. Balbuena, S. Vanildo, C.S. Catarina and A. Schevchenko et al., 2010. Twodimensional gel electrophoretic protein profile analysis during seed development of Ocotea catharinensis: A recalcitrant seed species. Braz. J.

Plant Physiol., 22: 23-33. DOI:10.1590/S1677- 04202010000100003.

Luo C, He XH, Chen H, Ou SJ, Gao MP. Analysis of diversity and relationships among mango varietys using Start Codon Targeted (SCoT) markers.Biochem Syst. Ecol. 2010; 38:1176-1184.

Manickavasagan, A., Essa, M. M. & Sukumar, E. (2012). Dates: production, processing, food, and medicinal values. CRC Press, pp.415.

Mann M, Ong SE, Grønborg M, Steen H, Jensen ON, Pandey A. 2001. Analysis of protein phosphorylation using mass spectrometry: Deciphering the phosphoproteome. Trends Biotechnol **20**: 261– 268.

Mrabet A.,Jimenezz-Araujo A.Fernández-Bola-os J., Rubio-SenentF., Lama-Mu-oz A., Sindic M.,et al. (2016). Antioxidant phenolicextracts obtained from secondary Tunisian date varieties (*Phoenixdactylifera* L.).Food Chem.196,917-924.10.1016.

Markhand GS, Abul-Soad AA, Mirbahar AA, Kanhar NA. 2010. Fruit characterization of Pakistani dates. Pak J Bot 42:37153722.

Marsafari M, Mehrabi AA. 2013. Molecular identification and genetic diversity of Iranian date palm (Phoenix dactylifera L.) varietys using ISSR and RAPD markers. Aust J Crop Sci 7:1160-1166.

Maryam, Jaskani MJ, Awan FS, Ahmad S, Khan IA. 2016. Development of molecular method for sex identification in date palm (Phoenix dactylifera L.) plantlets using novel sex-linked microsatellite markers. 3 Biotech 6: 22.

Mathew LS, Seidel MA, George B, Mathew S, Spannagl M, Haberer G, Torres MF, Al-Dous EK, Al-Azwani EK, Diboun I. 2015. A genome-wide survey of date palm varietys supports two major subpopulations in Phoenix dactylifera. G3 (Bethesda) 5: 1429-1438.

Meyer, W., T. G. Mitchell, E. Z. Freedman and R. Vilgays 1993) Hybridization probes for conventional DNA finger- printing used as single primers in the polymerase chain reac- tion to distinguish strains of Cryptococcus neoformans. J. Clin. Microbiol31, 2274-2280.

Mirbahar AA, Markhand GS, Khan S, Abul-Soad AA. 2014. Molecular characterization of some Pakistani date palm (Phoenix dactylifera L.) varietys by RAPD markers. Pak J Bot 46: 619-625.

Mitra C, Kharb P, Uppal S, Jain S (2011) Genetic diversity analysis in date palm (Phoenix dactylifera L.): a comparative assessment using ISSR and RAPD marker assays. J Hortic Sci Biotech 86: 398–402.

Mojtaba K, Mehdi R. Molecular markers in plants: Concepts and applications. 2015.

Moghaieb R.E.A, Abdel-Hadi A.A, Ahmed M.R.A, Hassan A.G.M. 2010. Genetic diversity and sex determination in date palms (Phoenix dactyliferaL.) based on DNA markers. Arab Journal of Biotechnology 13(2), 143-156.

Mrabet, A., Jiménez-Araujo, A., Fernández-Bola-os, J., Rubio-Senent, F., Lama-Mu-oz, A., Sindic, M., et al. (2016). Antioxidant phenolic extracts obtained from secondary Tunisian date varieties (Phoenix dactylifera L.) by hydrothermal treatments. Food Chem. 196, 917–924. doi: 10.1016/j.foodchem.2015.10.026.

Munshi. A and Osman, G. 2010. Investigation on molecular phylogeny of some date palm (Phoenix dactylifra L.) varietys by protein, RAPD and ISSRmarkers in Saudi Arabia. Aust. J. Crop Sci. 4(1): 23-28.

Okorley, E.L., J.A. Kwarteng and A.O. Mensah. 2005. Extension training needs in pesticides use by urban and peri-urban vegetable growers in central region of Ghana. J. Ext. Sys. 21:14-30.

Osman, A.M.A., and N.N. Boulos.(1978). Evaluation of some Sudanesedate varietys. ActaHortic: 185-190.

Osman AMA (1984). The performance of date palms in the Sudan. Acta Horticulturae 143: 231-237.

Ould Mohamed Salem A, Trifi M, Sakka H, Rhouma A, Marrakchi M (2001) Genetic inheritance analysis of four enzymatic systems in date palm (Phoenix dactylifera L.). Genet Resour Crop Ev 48: 361–368.

Pasqualone A, Lotti C, Bruno A,De Vita P, Di Fonzo N, Blanco A (2000)Use of ISSR markers for variety identification in durum wheat. Option Mediterranean"s, Série A, 40: 157–161.

Perezdela Torre, M.; Zirilli, P.; Ulrich, N.; Setten, L. and Escandon, A. (2010). Caracterización molecular en el género Mecardonia Ruiz & Pav.(Plantaginaceae) utilizando marcadores ISSR. Revista de la Facultad de Agronomía, La Plata, vol. 109, no. 1, p. 23-30.

Pinker I, Ibraheem YM, Böhme MH (2009) Propagation of some date palm varietys by using tissue culture methods. Acta Hortic 839: 71-78.

Rani R, Kharb P, Uppal S, Dhillon S, Batra P, Chowdhury V. 2007. Genetic divergence analysis in date palm (Phoenix dactylifera L.) genotypes using RAPD markers. Haryana J Hortic Sci 36: 101-103.

Reddy, Pradeep M., Sarla, N. & Siddiq, E. Inter simple sequence repeat (ISSR) polymorphism and itsapplication in plantbreeding. *Euphytica* **128,**

9–17(2002)..

Rhouma S, Zehdi SA, Ould Mohamed Salem A, Rhouma A, Marrakchi M, Trifi M (2007) Genetic diversity in ecotypes of Tunisian date palm (Phoenix dactyliferaL.) assessed by AFLP markers. J Hortic Sci Biotech 82: 929–933.

Rhouma-Chatti S, Baraket G, Dakhlaoui-Dkhil S, ZehdiAzouzi S, Trifi M (2011) Molecular research on the genetic diversity of Tunisian date palm (Phoenix dactylifera L.) using the random amplified microsatellite polymorphism (RAMPO) and amplified fragment length polymorphism (AFLP) methods. Afr J Biotechnol 10: 10352–10365.

Ruas, P.M.; Ruas, C.F.; Rampim, L.; Carvalho, V.P.; Ruas, E.A. and Sera,

T. (2003).Genetic relationship in Coffea species and parentage determinationof interspecific hybrids using ISSR (Inter- Simple Sequence Repeat) markers. Genetics and Molecular Biology, vol. 26, no. 3, p. 319-327.

Sakka, H., S. Zehdi, A. O. M. Salem, A. Rhouma, M. Marrakchi and M. Trifi. 2004. Genetic polymorphism of plastid DNA in Tunisian date palm germplasm (Phoenix dactylifera L.) detected with PCR-RFLP. Genetic Resources and Crop Evolution 51: 479-487.

Sammour RH, Sharaf-el-din A. 1989. Qualitative study on seed proteins of Thymelaea hirsute populations. Phyton (Austria) 29 (1): 83-92.

Sarnighausen,E,Wurtz,V,Heintz,D,Van Dorsselaer, A.and Reski ,R,(2004) Mapping the Physcomitrella patens proteome. Photochemistry, **65,** 1589– 1607.

Sedra MH, Lashermes P, Trouslot P, Combes M, Hamon S (1998) identification and genetic diversity analysis of date palm (Phoenix dactylifera L.) varieties of Morocco using RAPD markers. Euphytica 103:75-82.

Semagn K, Bjornstad A, Ndjiondjop M. 2006. An overview of molecular marker methods for plants.Afr J Biotechnol5 (25).

Shamsi& Mazloumzadeh, M Shamsi, H Nezamabadi-Pour - Computers and electronics in agriculture, 2009 Evaluation of general-purpose lifters for the date harvest industry based on a fuzzy inference system.

Siddiq, M., Aleid, S. M., and Kader, A. A. (2013). Dates Postharvest Science, Processing Technology and Health Benefits, 1st Edn. New Delhi: WileyBlackwell.

Soumaya RC, Sarra C, Salwa ZA, Khaled C, Khaled S (2014) Molecular polymorphism and phylogenetic relationships within Tunisian date palm (Phoenix dactylifera L.): evidence of non-coding trnL-trnF regions of chloroplast DNAs. Sci Hort 170:32–38.

Staub, J.E., F.C. Serquen, and M. Gupta. 1996. Genetic markers, map construction, and their application in plant breeding. HortScience 31:729– 741.
Thanh, V.C., Phuong, P.V., Uyen, P.H. and Hien, P.P. (2006): Application of protein electrophoresis SDS-PAGE to evaluate genetic purity and diversity of several varieties. Proceedings of International Workshop on Biotechnology in Agriculture.Nong Lam University Ho Chi Minh City, October 20-21, 192-194.
Tiwari, B. K., Brunton, N. P., and Brennan, C. S. (2013).Handbook of PlantFood Phytochemicals Source, Stability and Extraction, 1st Edn. Puducherry: Wiley-Blackwell.
Trifi M, Rhouma A, Marrakchi M (2000) Phylogenetic relationships inTunisian date palm (Phoenix dactyliferaL.) germplasm collection using DNA amplification fingerprinting. Agronomie 20: 665–671.
Varshney RK, Graner A, Sorrells ME. Genic microsatellite markers inplants: features and applications. Trends Biotechnol.2005; 23:48-55.
Vaz, AC, Pineheiro C, Martins JMN, Ricardo CPP (2004). Variety discrimination of Portuguese Lupinus albus by seed protein electrophoresis.The importance of considering "gutelns" and glycoproteins
.Field Crops RES.87 (1)23-34.

Virk, P. S., I. Zhu, H. J. Newbury, G. J. Bryan, M. T. Jackson and B. Y. Ford-Lloyd, (2000) Effectiveness of different classes of molecular markerfor classifying and revealing variation in rice (Oryza sativa) germplasm. Euphytica 112, 275-284.
Welsh J, McClelland M. Fingerprinting genomes using PCR with arbitrary primers. Nucleic Acids Res. 1990; 18:7213-7218.

Wessel D, Fluegge UI (1984) A method for the quantitative recovery of protein in dilute solution in the presence of detergent. Anal Biochem 138: 141–143.
Williams, J.F. (1990) Optimization strategies for the polymerase chain reaction. Biotechniques 7, 762–9. Wilson K and Walker J (2005). Principles and Techniques of Biochemistry and Molecular Biology. Cambridge University Press, Cambridge.
Wilson K and Walker J (2005). Principles and Techniques of Biochemistry and Molecular Biology. Cambridge University Press, Cambridge.
Wu, K.,R. Jones,L. Dannaeberger and P.A. Scolnik (1994) Detection of microsatellite polymorphisms without cloning.Nucleic. Acids.Res. 22, 3257-3258.
Xu,P.,Y. Zhang, L. Kang, M. J. Roossinck and K. S. Mysore. 2006. Computational estimation and experimental verification of off-target silencing

during posttranscriptional gene silencing in plants. Plant Physiol. 142:429- 440.

Yang M, Zhang X, Liu G, Yin Y, Chen K, Yun Q, Zhao D, and AlMssallemIS, Yu J. 2010. The complete chloroplast genome sequence of date palm (Phoenix dactylifera L.). PloS one 5: e12762.

Yang, B. Li, S.C. Zou, H.H.P. Fang, Zhang T Fate of antibiotic resistance genes in sewage treatment plant revealed by metagenomic approach WaterRes., 62 (2013), pp. 97-106.

Yusuf A, Culham A, Aljuhani W, Ataga C, Hamza A, Odewale J, Enaberue L. 2015. Genetic diversity of Nigerian date palm (Phoenix dactylifera) germplasm based on microsatellite markers. Int J Biosci Biotechnol 7: 121132.

Zaid, A. (2002).Date palm cultivation. FAO Plant Production and Protection Paper No. 156. Rome: Food and Agriculture Organization of the United Nations.

Zaid, A., & De wet, P. F. (2002). Botanical and systematic description of the Date Palm.In Z. A. (Ed.), Date Palm cultivation. Rev. Ed. Plant production and protection paper 156. Food and Agriculture Organization United Nations, Rome, pp. 73-105.

Zaid, A. and P.F. de Wet.(2002b). Date palm propagation, p. 73–105. In: A. Zaid (ed.). Date palm cultivation. Food and Agriculture Organization Plant Production and Protection Paper no. 156. Food and Agriculture Organizationof the United Nations, Rome, Italy.

Zehdi S, Trifi M, Ould Mohamed Salem A, Rhouma A, Marrakchi M(2002) Survey of inter simple sequence repeat polymorphisms in Tunisian date palms (Phoenix dactyliferaL.). J Genet Breed 56: 77–83.

Zehdi S, Trifi M, Billotte N et al (2004) Genetic diversity of Tunisian date palms (Phoenix dactylifera L.) revealed by nuclear microsatellite polymorphism. Heredity 141:278–287.

Zehdi S, Sakka H, Rhouma A, Salem AOM, Marrakchi M, Trifi M. 2004.Analysis of Tunisian date palm germplasm using simple sequencerepeat primers. Afr J Biotechnol 3: 215-219.

Zhao Y, Williams R, Prakash CS, He G (2012). Identification and characterization of gene-based SSR markers in date palm (Phoenixdactylifera L.). BMC Plant Biol 12: 237.

Zietkiewicz E, Richer C, Makalowski W, Jurka J, Labuda D (1994) a young Alu subfamily amplified independently in human and African great apes lineages. Nucleic Acids Res 22:5608–5612.

Appendixes

Appendix 3.1.1 Plant materials for protein and DNA analysis.

NO	Names	Abbr.
1	Khonaizi	M1
2	Sugay	M2
3	Khadhrawi	M3
4	Mishriq Wad Laggai	M4
5	Barhi	M5

Appendix 3.1.2 Reagents for protein analysis.

Stock (1):Acrylamide-bis acrylamide (30% T \2.6% C) stock solution: Acrylamide (30 grams) and 0.8 gram bis N.N. methylene bisacrylamidewere dissolved in 100 ml distilled water (each component dissolvedseparately). The solution was then filtered through glass filter under vacuumand kept at 4 °C in a dark bottle.

Stock (2):1.5 M Tris-HCl, pH 8.8:

Tris HCl (18.2 grams) was dissolved in 50ml distilled water. The pH was adjusted with HCl to 8.8 after that the volume was completed to 100 ml distilled water and the solution stored in the refrigerator.

Stock (3):0.5 M Tris-HCl, pH 6.8:

Tris-HCl (6 grams) dissolved in 50ml distilled water. The pH was then adjusted with HCl to 6.8 after that the volume was completed to 100 ml distilled water and the solution was stored in the refrigerator.

Ammonium persulfate solution (APS 10% W/V):

(freshly prepared) A solution of ammonium persulfate was prepared by dissolving 1.0 gram in 10ml distilled water, just before use in gel preparation.

Sodium dodecyl sulfate solution (SDS 10% W/V): Stock solution was prepared by dissolving 1.0 grams SDS in 10ml distilled water.

TEMED: Commercially available reagent was directly used.

Preparation of electrophoresis buffer (1X):

Dissolve 3.0 g Tris, 1g SDS and 14.1 g glycine in 1L of distilled water. pH adjustment is not essential.

Sample buffer:

Dissolve 5.0 ml (0.5 M Tris-HCl, pH 6.8), 0.5 g (SDS), 5.0 g (sucrose), 0.25ml (β-mercaptoethanol) (add before use), 5.0 ml (Bromophenol blue - 0.5% stock) in up to 50 ml with distilled water, and stored at 4°C for several months. Prior to use, it was warmed (37°C) and mixed briefly to completely dissolve SDS.

Protein staining solution:

Gel washing solution (GWS): 500 ml methanol and 100 ml Acetic Acid dissolved in 400 ml (distilled water).

Gel fixing solution (GFS): 500 ml Ethanol (95%) and 100 ml Acetic Acid: dissolved in 400 ml (distilled water).

Appendix 3.1.3 Reagents for DNA analysis.1M Tris HCl (pH 8.0), (100 ml). 15.76 g Tris-HCL was dissolved in 80 ml distilled water. pH was adjusted to

8.0 by adding concentrated HCl and final volume was adjusted to 100 ml. Solution was dispensed in to reagent bottle and sterilized by autoclaving. **0.5M EDTA (pH 8.0), (100 ml)**. 18.60 g EDTA di Sodium salt wasdissolved in 80 ml distilled water. pH was adjusted to 8.0 by adding NaOH pellets and final volume was adjusted to 100 ml. Solution was dispensed into reagent bottle and sterilized by autoclaving.

4M NaCl, (100 ml). 23.93 g NaCl was dissolved in 50 ml of distilled water and mixed well. When the salts get completely dissolved, final volume was adjusted to 100 ml. Solution was dispensed in to reagent bottle and sterilized by autoclaving.

70% Ethanol, (100 ml): 70 ml of ethanol mixed with 30 ml of distilled water. Dispensed in to reagent bottle and stored at 4°C.

Chloroform: Isoamyl alcohol (24:1), (100 ml): 96 ml of chloroform and 4 ml of isoamyl alcohol was mixed well and stored in to a reagent bottle at room temperature.

3M Sodium acetate (pH 7.0), (20 ml): 8.16 g Sodium acetate trihydrate salt (Merk) was dissolved in distilled water and the final volume was made up to20 ml. The pH was adjusted to 7.0.

Ethidium bromide (10mg/ml), (1 ml): 10 mg ethidium bromide was added to 1 ml of distilled water and it was kept on magnetic stirrer to ensure thatthe dye has dissolved completely. It was dispensed into an amber colored eppendorf tube and stored at 4°C.

1X TE buffer, (100 ml): 10mM Tris HCl (pH8.0) +0.1mM EDTA (pH 8.0):

1 ml of Tris HCl (1M), 200 μl of EDTA (0.5M) was taken and mixed with

98.8 ml of sterile distilled water. The solution was mixed thoroughly, autoclaved and stored at room temperature.

5X Tris Borate EDTA (TBE) buffer, (1000 ml) (pH 8.0)

54.5 g of Tris base, 27.5 g of Boric acid were taken, 20 ml of 0.5M EDTA (pH 8.0) was added. The final volume of 1 liter was adjusted by adding distilled water and the pH was adjusted to 8.0.

CTAB extraction buffer (3%), (10 ml): 1 ml of 1M Tris HCl (pH 8.0), 3.5 ml of 4M NaCl, and 0.4 ml of 0.5 M EDTA (pH 8.0) was measured and mixed with about 5 ml of hot distilled water, then 0.3 g (W/V) CTAB and 0.2 g (W/V) PVP (Sigma-Aldrich Inc., 3050 Supruce Street USA) wasadded to this. Dispensed in to reagent bottle.100 μl (1%) β-mercaptoethanol was added into mixture just before use.

TE-RNase solution: Commercially available reagent was directly used.

Agarose: Commercially available reagent was directly used.

Gel Loading Dye : Commercially available reagent was directly used. **100 bp DNA ladder**: Commercially available reagent was directly used.**Liquid N2:** Commercially available at liquid air industry.

Appendix 3.1.4 DNA extraction buffer.

1 ml of 1M Tris HCl (pH 8.0), 3.5 ml of 4M NaCl, and 0.4 ml of 0.5 M EDTA (pH 8.0) was measured and mixed with about 5 ml of hot distilled water, then 0.3 g (W/V) CTAB (Sigma-Aldrich Inc., 3050 Supruce Strect USA) and 0.2 g (W/V) PVP (Sigma-Aldrich Inc., 3050 Supruce Street USA)was added to this. Dispensed in to reagent bottle.100 μl (1%) β- mercaptoethanol was added into mixture just before use.

Appendix 3.1.5 DNA extraction glassware.

Balance, Freezer ,Refrigerated centrifuge, shaker water bath, pH meter, UV spectrophotometer, Thermal cycler, tube, mortar and pestle, magnetic stirrer and bar ,UV transilluminator, Hot air oven, incubator, Gel electrophoresis unit, mini centrifuge, shaker, laminar air flow.

Appendix 3.3.2 Preparation of polyacrylamide gel.Preparation of separating gel (12.5%):

Acrylamide-bisacrylamide (30%\2.6%)	6.67 mL
1.5 M tris/HCl buffer pH 8.8	4 mL
distilled water	5.05 mL
10% SDS	0.16 mL
Freshly prepared (10% amm persulphate)	0.12 mL
TEMED	0.008 mL

Preparation of stacking gel (7.5 %):

Acrylamide-bisacrylamide (30%\2.6%)	0.95 mL
1.5 M tris/HCl buffer pH 6.8	1.25 mL
distilled water	2.72 mL
10% SDS	0.05 mL
Freshly prepared (10% amm persulphate)	0.025 mL
TEMED	0.025 mL

Appendix 3.3.3. Protein staining solutions:

Coomassie brilliant blue (CBB-R250) :

Dissolve 0.25 g Coomassie brilliant blue (CBB-R250) in 200 ml (40%methanol), filter solution through Whatman No. 1 filter paper; add 200 ml (20% glacial acetic acid). Dye first dissolved in 50% methanol and then acetic acid was added.

Staining solution: 0.5 gm Commassie R-250 brilliant blue was dissolved in 250 ml methanol, 50 ml glacial acetic acid, completed by distilled water to 500ml mixed and kept at room temperature. Procedure Gels were stained overnight in 200 ml of Commassie brilliant blue R-250 solution.

Destaining solution: Destaining of protein was performed in 200 ml of destaining solution which composed of 250 ml methanol, 50 ml glacial acetic acid and 200 ml distilled water with gentle shaking. The destaining solution was changed several times until background colour was removed.

Appendix 3.4.1 DNA isolation: (CTAB protocol)

Cetyl trimethyl ammonium bromide (CTAB) method (D Coyle and Doyle,1987) 50-mg samples of young leaflets tissues were ground to a fine powder in liquid nitrogen, powder was then placed in 1.5-mL microtubes containing 700 μL 3% CTAB extraction buffer.

The solution was incubated at 65°C for 60 min, gently mixing by inversion every 15 min; an equal volume of chloroform-isoamylalcohol (24:1) was added to the tubes and gently mixed for 1 min.
Samples were centrifuged for 10 min. at 10,000 rpm; 600 μL of the supernatant was then transferred to a fresh tube following the addition of 500 μL chloroform-isoamylalcohol (24:1); this procedure was repeated twice; 500 μL of the supernatant was then transferred to a fresh tube with 700 μLof cold isopropanol (-20°C) and 1/10th volume of 3 M sodium acetate samples were gently mixed by inversion and centrifuged at 12,000 rpm for10 min, and so it was possible to visualize the DNA adhered to the bottom of the tube.
the liquid solution was then released and the DNA pellet washed with 700 μL of 70% ethanol, and set to dry for approximately 12 h, or until the next day, with the tubes inverted over a filter paper, at room temperature;
the pellet was then ressuspended in 100 μL TE buffer plus 5 mL ribonuclease (RNAse 10 mg mL–1) in each tube; this solution was incubatedat 37°C for 1h, and after stored at -20°C.They were used as template DNA for ISSR primer analysis.

www.ingramcontent.com/pod-product-compliance
Lightning Source LLC
LaVergne TN
LVHW051130210926
848482LV00001B/38
* 9 7 9 8 8 8 6 7 6 1 1 6 0 *